AF646706

RECENT TRENDS AND FUTURE PROSPECTS IN FOOD AND HEALTH SECURITY

ABOUT THE EDITORS

Prof. Shahnaz Jamil

Prof. Jamil is a Professor and Head, of the University Department of Botany, L.N. Mithila University, Darbhanga. She received her M.Sc. (1980) and Ph.D. (1998) from L.N. Mithila University, Darbhanga. She has successfully guided 05 Ph.D. students and published 20 research articles in noted and international journals. She has visited several countries such as Bangkok (Thailand), France, and Dubai (UAE) to present her papers. She has been a member of several academic societies such as ISCA, IBS, IPS and MBSI.

Dr. Gajendra Prasad

Dr. Prasad is an Associate Professor, University Dept. of Botany, L. N. Mithila University, Darbhanga, India. He has been highly proficient in Mycotoxicology. Earlier he worked as Senior Research Fellow (SRF) and Research Associate (RA) of CSIR, New Delhi in the field of mycotoxicology and he has already submitted FTR regarding SRF, RA and PL-480 projects in T.M. Bhagalpur University, Bhagalpur. He has published 40 research publications in reputed, international and UGC-CARE-listed journals, 9 book chapters and 3 review articles to his credit in Mycotoxicology. He has actively participated in the mycotoxin works and has delivered several lectures at international conferences and seminars in this field. He has already supervised 4 Ph.D. students successfully and is currently guiding 5 research scholars. He is life member of various academic bodies viz. ISCA, IBS, IPS and MBSI. He has over thirty years of teaching experience.

Dr. Khwaja Salahuddin

Dr. Salahuddin is an Assistant professor at the University Dept. of Botany, L.N. Mithila University, Darbhanga, India. He has been highly efficient in conducting microbiology, limnology, ethnobotany, and mycotoxicology research. He has been JRF and SRF of CSIR, New Delhi. He has published twenty research articles in noted, International and UGC-CARE-listed journals, one book chapter and one book (EXCLUSIVE BOTANY FOR NEET) to his credit. He has attended and delivered several lectures at several international conferences and seminars. He has supervised one Ph.D. student successfully and is currently guiding two research scholars. He is a life member of several academic bodies. He has fifteen years of teaching experience overall.

Dr. Ankit Kumar Singh

Dr. Singh is an Assistant Professor at L.N. Mithila University, Darbhanga, India. He obtained his doctorate from the Dept. of Botany, BHU in 2019. He has published 15 papers and 2 book chapters in reputed international journals with a high impact factor. Currently, he is editing 2 books with reputed international publisher, WILEY and 1 with Springer Nature.

RECENT TRENDS AND FUTURE PROSPECTS IN FOOD AND HEALTH SECURITY

EDITORS

Prof. Shahnaz Jamil

Dr. Gajendra Prasad

Dr. Khwaja Salahuddin

Dr. Ankit Kumar Singh

ACADEMIC ENCLAVE

© EDITORS

All rights reserved. No part of this publication may be reproduced, stored in a retrieval system, or transmitted, in any form or by any means, electronic, mechanical, photocopying, recording, or otherwise, without the prior permission of the publisher.

First Edition, 2025

ISBN: 978-93-48642-70-7 (Hardback)

Published by
ACADEMIC ENCLAVE
House No. 10/15, Ground Floor,
Block-A, Street No. 14/3, Kamal Vihar,
Kamal Pur, Burari, Delhi-110084
Email: info@academicenclave.com
Mobile: +91-7042423166
Website: www.academicenclave.com

Digitally Printed at Replika Press Pvt Ltd.

Cover Image Credit: www.foodsafety.gov

ललित नारायण मिथिला विश्वविद्यालय
LALIT NARAYAN MITHILA UNIVERSITY
कामेश्वरनगर, दरभंगा - 846004 बिहार
Kameshwaranagar, Darbhanga 846004 Bihar

प्रोफेसर संजय कुमार चौधरी
कुलपति
Prof. Sanjay Kumar Choudhary
Vice-Chancellor

Ref. No. VCR-209/25
Date 04:01:2025

FOREWORD

Food security is a global concern. It has become an alarming situation in developing nations where the human population is burgeoning day by day. In fact, food security ensures that people have physical, social, and economic access to safe, sufficient, and nutritious food that meets their dietary needs to lead healthy and active lives. There is no denying the fact that several measures have been undertaken by the Government, but the situation remains at a standstill. Food insecurity will cause and aggravate the health insecurity in India.

The researchers and academics are trying hard to address the issue of food security from all possible angles but in vain. There are so many cutting-edge tools and techniques of biotechnology for adequate food production along with clean water pollution-free environment and medicine. Over the last several scientists and academic missions have dwelt at length on food and health security which are a global concern as we know that without mitigating the problem of food and health insecurity the population of the world cannot be in equilibrium. The production of high-yielding and disease-resistant varieties of cereals and the latest transgenic plants with the help of genetic manipulations are the rays of hope in alleviating food insecurity. It will cause the problems of hunger, poverty, malnutrition and economic crisis if the problem of food security is not addressed on a priority basis because the population of India will be gigantic by 2050.

There are several neglected and underutilized crops especially millets which are required to be cultivated properly because these millets need minimal irrigation and withstand climate stress, insect pest attack and diseases. The food crisis is plagued with conflicts economic slow-down climate change and increasing prices of fertilizer. A multi-pronged strategy is the need of the hour to sustain food production and health security in all hemispheres. Holistic approaches should be adopted to mitigate the problems of food and health security. A deliberation by renowned academics and scientists is thus very welcome.

I would like to thank Professor Jamil, Head and her team of editors for compiling the academic outcomes that have resulted in this indispensable publication. I also thank the publisher, Academic Enclave, New Delhi, for bringing the proceedings in such a nice form.

Sanjay Kumar Choudhary

Contact : 06272 – 222463 (T-F) - Office, Mob.: +91 8544513251, E-mail vc@lnmu.ac.in, vc-lnmu-bih@nic.in

Preface

Food security is all about the use of safe and nutritious food by the people. It all depends on food availability, access, utilization, stability and sustainability. It is quite indispensable because the human population is increasing geometrically throughout the world. It is directly related to the economic development and poverty alleviation programme. Food security is a global challenge because the world has seen climate change, water scarcity and traditional farming, and disparity of food grain production. The issue of starvation and malnutrition is on the rise in some parts of the world. The time has come to rethink and revisit all the policies pertaining to food security which is directly correlated to the health security of the people.

Our country has made some progress in food production and distribution in recent years but still challenges are there. In 2022, the global food security index ranked India at 68th position out of the 113 major countries but in 2023 India ranked at 100th position out of 127 countries in the global hunger index. Information was collected at the international conference entitled"Recent Trends and Future Prospects in Food and Health Security" which was organised by the University Department of Botany, L.N. Mithila University, Darbhanga, Bihar (India) on May 29-30, 2023. The present volume is a collection of papers presented during this conference by the scientists and academics. The book contains 20 research papers and review articles by invited personalities focusing mainly on food and health security.

We are grateful to our mentors, colleagues and students for their continuous and valuable help in bringing about the book. We wish to thank all the authors who have not only participated in the international conference but also contributed to the book in the form of articles. We are obliged to the present Vice-Chancellor, Prof. S.K. Chaudhary who has written a foreword for this book and former Head and Dean, Prof. I.N. Mishra who guided me through everything to publish the book.

We hope that the book will cater to the needs of students, research scholars and teachers of plant science and agriculture. We are thankful to the publisher, Academic Enclave, New Delhi for bringing about the book in a nice format without any further delay.

Editors

Trends and
and Health Security

Contents

1

Evergreen Revolution in the Management of Food Security

Nand Kishore Sah

Former Vice-Chancellor, Veer Kunwar Singh University, Ara, Bhojpur-802301, Bihar, India.

INTRODUCTION

Food and health care management for the teeming millions living all over the world is very important for human resource development. By 2025, the world population is expected to cross 8.1 billion, as per the current UN report (1). The requirement of food grain by then may go up beyond 3 billion tons. By 2050, the world population is understood to reach about 9.7 billion, and proportionately the requirement of food will also rise. Scientific research in the last sixty years ushered in the 'Green Revolution' that fulfilled the adequacy of food in the world till the last century. Numerous systems, such as innovative farm implements, protection, healthy treated seeds, chemical fertilizers, bio-fertilizers, disease management measures using insecticides and pesticides, and the development of disease-resistant and climate-resilient plants have contributed a great deal in achieving this objective. However, this success has brought in numerous problems also which are detrimental to human health, soil fertility, the ecosystem, flavour of food, particularly vegetables, and so on.

There are certain microbes that affect production by over 15%. In some crops, as in potato, the level of reduction in the yield may be as high as 30% (2). Certain viral diseases may also reduce the yield by about 3%. Several factors including the edaphic and other climatic conditions have been shown to influence the above-mentioned pathological conditions.

Modern scientists' attention has, therefore, shifted from just food production to food production without compromising the quality of human life and the environment. Professor M. S. Swaminathan, the chief architect of the first successful 'Green Revolution' in India, has propounded a novel concept of 'Evergreen Revolution' that takes care of most of the obnoxious issues that emerged from the first 'Green Revolution'. Biotechnology holds the key to solutions for achieving the goals of the 'Evergreen Revolution' (3).

Impact Factors that may Influence 'Evergreen Revolution'

There are many factors which may help us overcome the impediments in crop management for boosting food production. For example, soil management, discouraging mono-cropping, plant disease management proper use of green fertilizers, raising climate-resilient and resistant varieties, identification and use of additional crop plants from the biodiversity pool are the identified areas for accelerated research for achieving the goals of the 'Evergreen Revolution. The ancient Indian Rishis had reported existence of about 8.4 million living species on this planet. The modern scientists have so far recognized about 1.7 million of them. Still some 6.7 million remain to be identified. It is quite possible that some of the species in the biodiversity pool may fulfill the criteria of being listed as favorite crop plants (4).

Disease management of the crop plants is very important for meeting the production targets. The first 'Green Revolution' succeeded primarily owing to measures including disease management of the crop plants. Six major principles have been identified in this pursuit: Exclusion, Eradication, Protection, Resistance, Treatment/Therapy, control of vectors, such as, Insect and Weed.

Exclusion: This principle works on the famous precept 'Prevention is better than cure'. Here, entrance of pathogens in the crop field is minimized and restricted largely to a marked area only. This principle holds great promise as it helps avoid the use of pesticides, insecticides, weedicides, nematocides and so on that contribute to environmental pollution leading to health risk. Crop rotation and management is important in achieving this goal.

Eradication: This principle ensures complete removal of the potent threats including the infected crop plants. It applies to individual plant, seed lot in a limited area/field/region soon after introduction. Once established it is difficult to eradicate a pathogen. For a larger area, generally it does not work satisfactorily. These problems may be addressed through good sanitation, crop rotation, burning of infected plants, pasteurization, fumigation and solarisation of the soil, disinfection of irrigation water by ultraviolet irradiation, chlorination, bromination and ozonation (5).

Protection: This practice of crop protection has been in practice ever since human civilization adopted agriculture as a primary means of livelihood. Protection then was largely from the herbivores and rodents. Later, certain diseases were also identified. Till 1940, most of these diseases were managed with the use of Bordeaux mixture made of inorganic components {Copper sulphate, (CuSO4) and slaked lime i.e. calcium hydroxide, Ca(OH)2}. Several other agrochemicals including the organic chemicals were included in the pesti/fingicides regime, which enhanced crop production. Over the years, however, the pathogens developed resistance against several agrochemicals. They have now become a knotty challenge for the agricultural scientists to keep the pace of production satisfactory while minimizing environmental pollution and health risk. For this, it is imperative that the use of pesticides and fungicides is minimized without compromising crop production (Berlin et al, 2018). Multi-cropping and microbe-soil interface are also significant areas that need to be considered for the sustainable boost to production (7).

Resistance: Developing crop plants resistant to several types of pathogens is desirable. Modern biotechnological tools and techniques have led us to a new era of crop disease management. Precise knowledge of the disease and genetic engineering coupled with tissue culture based resistance to the crop plants have greatly reduced the time of evolving new varieties. Several types of pathological conditions may now be overcome without serious complications.

Treatment/Therapy: The crop plants that have already developed pathological conditions due to pathogens, deficiency of required elements in the soil, hostile climatic conditions and other reasons are treated with inorganic, organic and bio-molecules to render them healthy again. A plethora of such chemical molecules have been investigated that help overcome the pathological conditions partially and sometimes fully. Some of these molecules enter food chain and have been found to disturb environmental harmony leading to unpleasant issues. The principle of crop rotation and cropping system may be followed to overcome the problem of pathogenicity to a large extent.

Control of vectors: The vectors of pathogens, such as bacteria, fungi, viruses, mycoplasma are often air, water, infected seeds and plants, birds, rodents, lepidodepterons insects, mollusks, nematodes and farm implements among others. Their complete elimination is not always possible. For a limited region it is possible, but for the vast tracts of cultivated lands it is extremely difficult. These efforts may support production, but may also create ecological problems associated with biodiversity.

MAJOR BIOTECHNOLOGICAL TOOLS AND TECHNIQUES

Three primary processes are followed in biotechnology i.e. (a) Genetic Engineering or Recombinant DNA Technology (b) Tissue Culture (c) Transgenics.

(a) Genetic Engineering comprises synthesis of genes or part thereof, repair of genes, combining genes isolated from one organism/plant with others (Recombinant DNA), manipulation of the artificial genes or the parts thereof with the recombinant DNA with an aim to improve microbes, plants, animals and even humans. Genetic engineering entails two basic processes: (i) Identification, Synthesis or Isolation/ Cutting of the desired gene or part thereof and (ii) pasting them at suitable sites. The genes of interest may be cloned by the process of amplification (synthesis) or cutting them out from the host sites using precise biochemical scissors known as restriction endonucleases which cleave phosphodiester bonds at defined sites (Nucleotide sequences). These are then inserted into compatible plasmid DNA which act as vehicles to transport the gene(s) to the target plants/organisms. These techniques have been perfected now. The credit for cloning a gene for the first time goes to Cohen and Boyer (1973) (6) who succeeded inserting a foreign DNA fragment into a plasmid of *Escherichia coli*. Plasmids are normally circular DNA molecules with replication ability independent of chromosomal DNA found in certain bacteria. Paul Berg in 1979 was possibly the first to introduce a gene of Simian virus 40 in bacteria through lambda phage, and therefore, known as 'Father of genetic engineering'. He was awarded Nobel prize in 1980 with Gilbert and Stanley. Plant biotechnology got a shot in the arm since *Agrobacterium tumefaciens* was recognized as a sound vehicle to transfer foreign gene (s) into plants, particularly of the dicotyledonous group. This bacterium hosts versatile Ti-plasmids that carry foreign genes into the plant cells during the process of transformation. *A. tumefaciens* is, therefore, said to be a natural genetic engineer.

(b) **Tissue Culture:** The transformed cells are developed as callus tissue which eventually gives rise to plants carrying the cloned foreign genes. Tissue culture is a versatile *in vitro* method of propagation of plants with desired traits. It is now almost a perfected technique. Major factors that influence tissue culture include Genotype and physiological condition of plants, Culture media, Culture conditions and Hormones. Research in the agricultural sciences have got a shot in the arm since the introduction of tissue culture techniques. Not only with regard to food production, but also in the arena of horticulture, quality improvement, pharmaceutics and environmental safety plant tissue culture has contributed significantly. Plants and their parts are now being used as factories for mass production of necessary secondary metabolites, pharmaceuticals, nutraceuticals and so on. Plant tissue culture has contributed significantly in the development of disease resistant, disease-free and climate resilient crop plants (8).

(c) **Transgenics:** The swelling global population, climate change, genetic erosion and increasing need for nutritional supplements have necessitated the development of new cultivars adaptable to different biomes with better resistance to biotic and abiotic stresses for improved yield and quality, and thus encounter the future challenges (9-11). The new biotechnological tools and methods may equip the elite crop plants with smart features including quantitative and qualitative supremacy over the conventional counterparts. The biochemical tool, CRISPR-Cas9 has enabled the scientists to develop choice crop plants with the desired traits.

Current Biotechnological methods for viral disease resistance in the crop plants: Multilayered defenses against microbial pathogens have recently been applied through the versatile tools of genetic engineering, particularly against the viral pathogens. These objectives have been realized either by cloning trans-resistance genes or by silencing the disease-sensitive genes by various sound methods. RNA interference (RNAi) system has been deployed in the suitable plants for anti-viral defense through transgenic expression of the plus or minus strand of the viral genome, inverted repeats or artificial microRNAs (amiRNAs). CRISPR-Cas {(Clustered Regularly Interspaced Short Palindromic Repeats)-(CRISPR-Associated Systems or Proteins)} has been used to gain resistance against the DNA and RNA viruses. Broad resistance has been conferred on several crop plants simply by stacking several resistance genes. RenSeq is another powerful tool of the comparative genomics which makes use of nucleotide-binding leucine-rich repeats (NLR) gene family members having proven role in strengthening disease resistance. The NLR-gene specific DNA sequences are used as baits to fish out such sequences from the resistant germplasms. The enriched NLR-DNAs are sequenced and compared with the reference genomes to identify the polymorphisms in the NLR-genes for jacking up resistance. Inactivation of susceptibility inducing genes is also a ploy to boost resistance (2).

Diseases overcome through Biotechnology: *Bacillus thuringinensis* toxin gene, commonly known as Bt gene, was one of the first to be cloned into cotton that effectively controls damage caused to the capsule of the plant by the caterpillars of several pests, particularly of budworm i.e. *Heliothis virescens* (F.), and bollworm i.e. *Helicoverpa zea* (Boddie). It has tremendous pecuniary and industrial significance. Bt genes encode toxic proteins, commonly known as Cry-proteins, that appear crystal-like in the carrier plant cells. More than sixty types of Cry-proteins have been identified which may show varying degrees of protective effects against the pests.

How does it work? When the insects feed upon these plants, the crystals of Cry proteins enter the gut of insects and get activated by their own digestive enzymes. The active Cry proteins reach the lining cells of the alimentary canal through the specific receptors and rupture them. The insects then either die or stop feeding on the plant. Several crop plants have been protected by invoking this and other strategies (Table 1). However, the concerns about the human use of such crops are quite genuine. In India, except Bt-cotton many such crops are under scanners of the government machinery.

The first viral resistant crop plant was developed in 1986 by Abel and coworkers by transferring TMV viral coat protein to the tobacco plant. Several crop plants have been raised that confer resistance against various viral diseases (Table 2). CRISPR-Cas mechanism is applied these days quite often to enhance the level of resistance against the viral diseases in the crop plants, although the sound results of the field trials are still awaited.

Table 1: Genetically modified crop plants that resist bacterial/fungal pathogens:

Sl. No.	Name of the Cereal	Pathogen(s)	Biotechnological intervention	Genes involved
1.	Wheat	*Pseudomonas syringae pv. oryzae*	Interspecies/interfamily transfer of known *PRRs	AtEFR
2.	Wheat	*Blumeria graminis f. sp. tritici*	Interspecies/interfamily transfer of known PRRs	TaRLK1 and TaRLK2
3.	Wheat and Barley	*Blumeria graminis f.sp. hordei; Blumeria graminis f.sp. tritici*	Interspecies/interfamily transfer of known PRRs	HvLEMK1
4.	Wheat	Blumeria graminis f. sp. tritici	Interspecies/interfamily transfer of known PRRs	HvLecRK-V
5.	Wheat	*Bipolaris sorokiniana*	Altered expression of transcription factors for better signalling	TaPIMP1
6.	Wheat	*Fusarium graminearum*	Host-induced gene Silencing by RNAi	FgCh3b
7.	Wheat	*Fusarium culmorum*	Silencing of host Genes by CRISPR-Cas9	FcGls
8.	Wheat	*Blumeria graminis f. sp.tritici*	Silencing of host Genes by CRISPR-Cas9	TaMlo-A1
9.	Wheat	*Blumeria graminis f. sp.* tritici	Silencing of host Genes by CRISPR-Cas9	TaEDR1
10.	Wheat	*Fusarium graminearum*	CRISPR-Cas9	TaLpx-1
11.	Wheat	*Fusarium graminearum*	CRISPR-Cas9	TaHRC

Sl. No.	Name of the Cereal	Pathogen(s)	Biotechnological intervention	Genes involved
12	Wheat	*Puccinia triticina, P.* graminis and P. striiformis	Host-induced gene Silencing by RNAi	PtMAPK1, PtCYC1, PtCNB
13.	Durum wheat, Barely, Rice, Sorghum, Maize	*Multiple biotrophic pathogens*	Transfer of APR alleles	TaLr34
14.	Barley	Multiple rusts and powdery mildew	Transfer of APR alleles to boost resistance	TaLr67
15.	Barley	*Fusarium graminearum*	Host-induced gene Silencing by RNAi	FgCYP51A, FgCYP51B, FgCYP51C
16.	Rice	*Xanthomonas oryzae pv.* oryzae-derived elf18	Interspecies/interfamily transfer of known PRRs	AtEFR
17.	Rice	*Acidovorax avenae subsp.* avenae	Interspecies/interfamily transfer of known PRRs	AtEFR
18.	Rice	*Pseudomonas syringae pv. tomato; Agrobacterium tumefaciens; Xanthomonas oryzae pv. oryzae*	Production of chimeric receptor kinases and R genes	AtEFR-OsXa21
19.	Rice	*Magnaporthe oryzae*	Production of chimeric receptor kinases and R genes	OsXa21-OsCEPiP

Sl. No.	Name of the Cereal	Pathogen(s)	Biotechnological intervention	Genes involved
20.	Rice	*Xanthomonas oryzae pv. oryzae*	Deletion of effector binding sites	Os11N3/OsSWEET14
21.	Rice	*Xanthomonas oryzae pv. oryzae*	Addition of effector binding sites	OsXa27
22.	Rice	*Broad-spectrum of pathogens*	Altered expression for signaling components	AtNPR1
23.	Rice	*Magnaporthe oryzae*	Transfer of APR alleles to boost resistance	OsIPA1/OsSPL14
24.	Rice	*Magnaporthe oryzae*	CRISPR-Cas9	OsSEC3A
25.	Rice	*Magnaporthe oryzae*	CRISPR-Cas9	OsERF922

*PPRs : Pattern-Recognition Receptors

Table 2: Genetically engineered certain crop plants which resist viral pathogens

Sl. No.	Name of the crop	Type/Name of the virus	Level of biotechnological intervention	Genes/ Protein involved
1.	Potato	Potato virus Y	*A. tumefaciens* mediated viral coat protein tranfer	Coat protein
2.	Potato	Potato leaf roll virus	*A. tumefaciens* mediated viral gene tranfer	Replicase and Helicase
3.	Papaya	Papaya ring spot virus	*A. tumefaciens* mediated viral coat protein tranfer	Coat protein
4.	Squash	Watermelon mosaic virus 2, Zucchini yellow mosaic virus	*A. tumefaciens* mediated viral coat protein tranfer	Coat protein
5.	Sweet pepper	Cucumber mosaic virus	*A. tumefaciens* mediated viral coat protein tranfer	Coat protein
6.	Tomato	Cucumber mosaic virus	*A. tumefaciens* mediated viral coat protein tranfer	Coat protein
7.	Plum	Plum pox virus	*A. tumefaciens* mediated viral coat protein tranfer	Coat protein
8.	Bean	Bean golden mosaic virus	RNAi-mediated resistance	+, - RNA of viral replication protein

CONCLUSION

The world population is expected to exceed 9.7 billion by 2050, albeit the land and other resources on earth remain limited. New strategies are needed, therefore, to guarantee food security and better future of mankind. Novel efforts are required to exploit the 75 % of the earth's surface which is spanned by water. We also need to delve deep into the treasure of biodiversity for identification of novel genes/species to meet the future challenges of food security. Feeding over 8 billion people and managing other requirements including clean water, health, quality life and tidy environment is indeed a great challenge. Remarkable progress has been made in achieving the goal of food adequacy in the world. Management of diseases of the crop plants is a sound step towards this mission. During the first green revolution food production was optimized through the chemical and biological control of diseases of the crop plants. These achievements unfortunately came along with numerous obnoxious issues, particularly related to health and environment. Ideal management of crop diseases for food security is possible through the principles of 'Evergreen Revolution' which makes use of the innovative ideas and the latest biotechnological tools and methods. Exclusion, Eradication, Protection, Resistance, Treatment/Therapy, control of vectors, such as, Insect and Weed are the recognized ways to manage crop diseases. Biotechnology has contributed a great deal via three successive processes: 1.

Genetic Engineering or Recombinant DNA Technology 2. Tissue Culture and 3. Transgenics. RNAi and RenSeq methods are good for raising crop varieties with reduced pathogenic attacks. CRISPR-Cas9 mediated genome editing has yielded remarkable success in developing choice crop plants bearing the desired traits.

REFERENCES

2. Oliver Xiaoou Dong and Pamela C. Ronald (2019) Genetic Engineering for Disease Resistance in Plants: Recent Progress and Future Perspectives. *Plant Physiol.* 180: 26–38. doi: 10.1104/pp.18.01224

3. Swaminathan M.S. (2010). From Green to Evergreen Revolution. New Delhi: Academic Foundation.

4. Sah N K (2018). Scope of Plant Biotechnology in the developing Countries. Plant Biotech. Ed. By B. D. Prasad, Sahni S, P. Kumar and Siddqui M. W. pp 45-66.

5. Ogle H. (2021). Disease management: Exclusion, Eradication and Elimination. Ed. *Plant Pathogens and Plant Diseases: Australasian Plant Pathology Society* by JF Brown and HJ Ogle. pp 358-72.

6. Akanmu A O, Babalola O O, Venturi V et al. (2021). Plant Disease Management: Leveraging on the Plant-Microbe-Soil Interface in the Biorational Use of Organic Amendments. *Front. Plant Sci.* https://doi.org/10.3389/fpls.2021.700507

7. Boyer & Cohen - GNN (1973) - Genetics and Genomics Timeline http://www.genomenewsnetwork.org.

8. Twaij, B M, Jazar, Z H and Hasan, M N (2020). Trends in the use of tissue culture, applications and future aspects. *Int. J. Plant Biol.* https://doi.org/10.4081/pb.2020.8385

9. Arzani, A., and Ashraf, M. (2017). Cultivated ancient wheats (Triticum spp.): a potential source of health-beneficial food products. *Comp. Rev. Food Sci. Food Saf.* 16, 477–488.

10. Basso M F , Monteiro F B A , Grossi-de-Sa M et al (2020). Insights Into Genetic and Molecular Elements for Transgenic Crop Development. *Front. Plant Sci.,* 15 May 2020 | https://doi.org/10.3389/fpls.2020.00509

11. Bigini V, Camerlengo F, Botticella E, Sestili F, Savatin D V (2021). Biotechnological Resources to Increase Disease-Resistance by Improving Plant Immunity: A Sustainable Approach to Save Cereal Crop Production. Plants, 10, 1146. https://doi.org/10.3390/plants10061146.

2

Isolation, Characterization and Identification of Endophytic Bacteria from Roots of *Hordeum vulgare* Plants

Aisha Parween and Vidyanath Jha

University Department of Biotechnology, L. N. Mithila University, Darbhanga, Bihar

INTRODUCTION

Endophytes are group of bacteria having the ability to enter inside the plant hosts colonizing the intercellular spaces. They exist in a range of tissues types within a broad range of plants, colonizing the plant systemically with bacterial colonies. They are ubiquitous, colonize most of the plants, and have been isolated from almost all plants examined till date. Endophytes are to live symbiotically within the plants. This type of association of endophytes with internal tissues of host plant has increasingly gained them scientific and commercial interest due to their potential to improve plant quality and growth (Carroll, 2001: Schulz *et al.* 2000).They exhibit complex interactions with their hosts which involves mutualism and antagonism.

Their association can be obligate or facultative. Plants strictly limit the growth of endophytes, and these endophytes use many mechanisms to gradually adapt to their living environments. In order to maintain stable symbiosis, endophytes produce several compounds that promote growth of plants and help them adapt better to the environment. Some of the endophytes are known to protect their host from being attacked by fungi, insect and mammals by producing secondary metabolites (Zhang, 2007). Among them, endophytic bacteria are thought to interact closely with their host plants, and therefore could be used as biological control agents in sustainable crop production potentially (Sturz and Nowak, 2000; Taechowisan et al., 2003; Zhang *et al.*, 2008).

Endophytes are known to supply nutrients to plant by fixing atmospheric nitrogen and solubilizing ion (Marx 2004; Porras- soriaano *et al.* 2009). This ultimately leads to increase in plant immune system as well as protects plant from infection by plant pathogens .studies have also shown role of endophytes in removal of soil contaminants (Barac *et al.* 2004; Doty *et al.* 2009). Studies on endophytes their significance and role in plant metabolism is an important area to explore.

The present studies was therefore undertaken to investigate and identify endophytic bacteria present in the tissue of roots barley plant. The technique of 16s rRNA partial sequencing is also used for identification of the bacterial species. Studies were also performed on biochemical characterization of the endophytes and their ability to interact with different antibiotic.

MATERIALS AND METHODS

Collection of the sample: Three months old plants of Barley were taken from a plant farm near Darbhanga, Bihar. These plants were maintained under controlled environmental conditions in well plastic pots in the green house of Department of Biotechnology, L.N. Mithila University, Darbhanga Bihar, India.

Isolation of endophytic bacteria from Barley roots

Surface sterilization: The root sample of the *Barley* plant were taken and processed separately. Healthy and undamaged root were collected from the plant grown under controlled environmental conditions. These were than washed under running water and dried followed by surface sterilization. Surface sterilization was done first washing in 1% savlon for 5-6 minutes. Subsequently these pieces of roots were treated with 0.1 % mercuric chloride for surface sterilization.

Isolation of endophytic Bacteria: The samples of roots were cut into small pieces and macerated separately in phosphate buffer of pH 7.2 with a sterile pestle and mortar. Tissue extract were then prepared for tenfold dilution in sterile saline. Serial dilutions were prepared from this extract. For inoculations 0.1ml of the aliquot was used on Nutrient Agar medium. The inoculations were done in triplicates separately for both roots tissue extract. These plates were then incubated at 37^0C. Observations were taken after 48 to 72 hrs .Bacterial colonies were differentiated on the basis of morphological colony characters. Bacterial isolates were picked from plates and Purified by streaking techniques and incubated at 37^0C. The isolation process repeated till monocultures were obtained for further experimentations .The media used for isolation was Yeast Manitol Agar, Picovasky's medium.

Characterization of endophytic bacteria

(a) Morphological: Morphological characterization was done by Gram staining and motility test.

(b) Physiological Characterization

Effect of pH on growth of isolates: Optimization and standardization of growth of the isolates on range of pH was studied by inoculating the isolates on Nutrient broth media (Hi-media) having pH range from 1 to 14. These were then incubated at 37^{0}C. for 48 hours. The growth was determined by taking optical density at 520 nm.

Effect of temperature on growth of isolates: Effect of temperature on growth of isolates was determined by inoculating the isolates in nutrient broth and Incubated at different temperature that is at 4^{0}C, 28^{0}C, 37^{0}C, 50^{0}Cfor 48 hours. Optical density at 520 nm was measured.

Biochemical Characterization

(a) **Catalase Test**: Catalase activity of the isolates was estimated by using H_2O_2 solution onto the microscopic slides containing the culture of the isolates separately.

(b) **Amylolytic activity:** Amylolytic activity was observed after inoculating the isolates in nutrient agar with 1% starch of pH 6.06. Culture plates were treated with iodine after incubation period. Clear zone was observed around colonies confirming the activity.

(c) **Urease activity:** Isolates were grown on the medium containing urea agar. After incubation slant were observed for change in its colour from redish pink indicating positive urease activity.

(d) **Antibiotic Sensitivity test:** The isolates to be screened for their sensitivity for antibiotics were cultured on to the nutrient agar plate. Antibiotic Kanamycin of varying concentration (5µg/ml, 50µg/ml, and 100µg/ml) used for treating the culture by using well technique. The protocol was repeated for Ampicillin and Amoxicillin antibiotic with concentration – 5µg/ml, 50µg/ml, and 100µg/ml on separate plates. Cultures inoculated for 24 hrs. at 37^{0}C the growth of the bacterial colonies was observed and zone of inhibition measured.

RESULTS AND DISCUSSION

Three bacterial colonies were isolated from root tissue extract. Two bacterial colonies isolate II& III were found to be of same species.

Table 1. Morphological characterization of endophytes

Colony characters	Isolate I/ root	Isolate II/ root	Isolate III/ root
Size	4mm	3mm	3.7mm
Shape	Circular	Circular	Circular
Colour	Orange	Yellowish	Yellowish
Margine	Regular	Regular	Regular

Colony characters	Isolate I/ root	Isolate II/ root	Isolate III/ root
Opacity	Opaque	Opaque	Opaque
Elevation	Flat	Flat	Flat
Consistency	Smooth	Smooth	Smooth
Gram characters	Gram positive	Gram negative	Gram negative
Motility	Motile	Motile	Motile

Table 2: Screening of the isolates for enzyme activity

Organism	*Amylolytic activity*	*Urease activity*	*Catalase activity*
Bacillus subtilis	+	+	+
Pseudomonas spp.	-	-	+

+ = Positive, - = Negative.

Graph: Effect of pH on growth of isolates

Media: Nutrient Brothincubation: temp 37 ^{0}C time: 2 Days

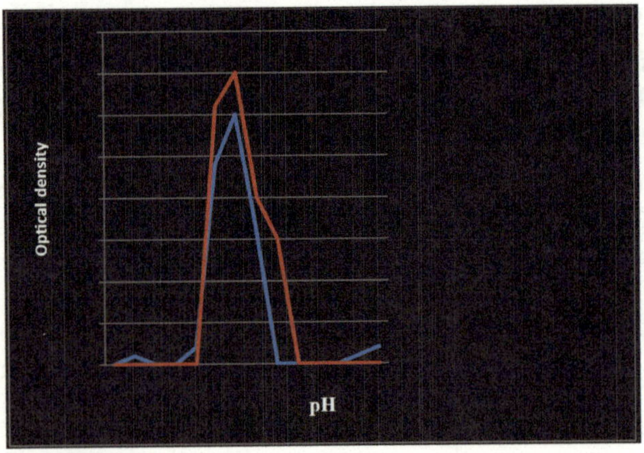

RESULTS AND DISCUSSION

Hordeum vulgare is used in many industries since many years. However, new discoveries in this plant research leads to development and opening of many more areas to explore. The plant is now gaining importance to

Develop some more new search for the future development by understanding the gene level studies. Therefore, considering its huge uses, there is an ample scope for future research and hence further pharmacological investigations are warranted. Endophytic bacteria are reported to enhance the growth of plant and may also increase the supply of nutrients to plants in turn helping in building the defence mechanism of the plant.

The present investigations were undertaken to find out the presence of bacterial endophytes in root of this plant. By using nutrient agar and selective medium (YMA and Picovasky's medium) two endophytes were isolated from roots tissue of the plant. Hence ultimately two endophyte were present in root tissue of the barley plant and further characterization and sequencing were performed in these two isolate.

On the basis of the morphological and biochemical characterization of the endophytes they are identified as *Bacillus subtilis* and *Pseudomonas spp.* The isolate I shows 99% identity to *Bacillus subtilis* strain. The isolate 2 shows 99% identity to *Pseudomonas spp.* Strain. . The sequences are submitted to NCBI GENE BANK and they are under process.

During present studies the endophytic bacteria *Bacillus subtilis* isolated from roots showed high amylolytic, catalase And urease activity though only catalase activity observed in *Pseudomonas* spp. (Table2). Sensitivity of Three different antibiotics at varying concentration was also tested for these two endophytes. *Bacillus subtilis* was Observed to be resistant to all three antibiotics kanamycin, ampicillin &amoxicillin as no zone of inhibition was observed whereas *Pseudomonas spp.* showed sensitivity to ampicillin at higher concentrations.

To conclude, two bacterial endophytes are isolated from root tissue of *Hordeum vulgare* plant. Both the Isolates *Bacillus subtilis* & *Pseudomonas spp.* were present in root tissue. Further research work is required to explore more about the role of these bacterial endophytes in the metabolism of the plant.

REFERENCES

1. Adachi, K., Nakatani, M. and Machida, H. (2002). Isolation of an Endophytic Diazotroph, *Klebsiella oxytoca*, from Sweet Potato Stems in Japan. Soil Science and Plant Nutrition., 48 (6): 889-895.
2. Ahmad, M. H. Rafique Uddin, M., McLaughlin, W. (1984). Characterisation of indigenous Rhizobia from wild legumes. FEMS Microbiology Letters, 24: 197 –203.
3. Ajit Varma, Savita Verma, Sudha, Nirmal Sahay, Britta Bu¨Tehorn, and Philipp Franken *Endophyte Applied And Environmental Microbiology,;*P. 2741–2744 June (1999)
4. Anonymous (1957). Manual of Microbiological Methods, McGraw Hill Book Company Inc. New York. P.127.
5. Anzai, Y. Kim, H., Park, J. Y., Wakabayashi, H. and Oyaizu, H. (2000). "Phylogenetic affiliation of the *Pseudomonas* based on 16S rRNA sequence". International Journal of Systematic Evolutionary Microbiology, 50 (4): 1563-89.
6. Arite Wolf, Antje Fritze, Martin Hagemann and Gabriele Berg; *International Journal of Systematic and Evolutionary Microbiology* (2002)

7. Banik A, Mukhopadhaya SK and Dangar TK (2016) Characterization of N2 -fixing plant growth promoting endophytic and epiphytic bacterial community of Indian cultivated and wild rice (*Oryza* spp.) genotypes, Planta. 45: 56-59.

8. Bathlomew JW (1962) Variables influencing results and precise definition steps in gram staining as a means of standardizing the results obtained. Stain Technol 37: 139-55.

9. Beever RE and Burns DJW (2000) Phosphorus uptake, storage and utilization by fungi. Advance Bot Res, 8: 127–39.

10. Cavalcante, V. A. and Dobereiner, J. (1988). A new acid-tolerant nitrogen-fixing bacterium associated with sugarcane. Plant Soil, 08: 23-31.

11. Chakmakci M. L., Evans H. J. and Shidler R. J. (1981). Characteristics of Nitrogen-Fixing *Klebsiella oxytoca* isolated from wheat roots. Plant and Soil, 61: 53-63.

12. Chan, Y. K. Barraquio, W. L. and Knowles, R. (1994). N_2-fixing Pseudomonads and related soil bacteria. FEMS Microbiology Reviews, 13: 95-118.

13. Chanway, C. P. (2002). Plant growth promotion by *Bacillus* and relatives, pp.219-235.

14. Chelius, M. K. and Triplett, E. W. (2000). Immunolocalization of dinitrogenase reductase produced by *Klebsiella pneumoniae* in association with Zea mays L. Applied and Environmental Microbiology, 66: 783-787.

15. Dhanya N. Nair and S. Padmavathy; Impact of Endophytic Microorganisms on Plants, Environment and Humans Hindawi Publishing Corporation Scientific World Journal, Article ID 250693, 11 pages Volume (2014)

16. Dudeja SS and Giri R. (2014). Beneficial properties, colonization, establishment and molecular diversity of endophytic bacteria in legumes and non-legumes. African Journal of Microbiology Research, 8 (15): 1562-1572.

17. Dunne C, Crowley JJ, Loccoz YM, Dowling DN, Bruijn FJ and O"Gara F. (1997) Biological control of *Pythium ultimum* by *Stenotrophomonas maltophilia* W81 is mediated by an extracellular proteolytic activity. Microbiol, 143: 3921-31.

18. Dworkin, M and Foster, J. (1958) Experiments with some microorganisms which utilize ethane and hydrogen. Journal of Bacteriology, 75: 592-601.

19. Garcia A., Rhoden S.A., Bernardi-Wenzel J., Orlandelli R. C., Azevedo J.L., Pamphile J.A. *Journal of Applied Pharmaceutical Science* Vol. 2 (10), pp. 035-040, October, (2012)

20. Hui Li, Xiaoxian Wang, Meizhe Han, Zhenzhen Zhao, Minqian Wang, Qin Tang, Changhong Liu, Brian Kemp, Yucheng Gu, Jinglei Shuang and YarongXue; *African Journal of Biotechnology* Vol. 11(1), pp. 231-242, 3 January, (2012)

21. Irene De Araújo Barros; Welingtonluiz Araújo; Joãolúcio Azevedo; *Brazilian Journal Of Microbiology* Issn 1517-8382 41: 956-965 (2010)

22. Jema Haji and Abdu Mohammed *Ethiopia African Journal of Agricultural Research* Vol. 8(9), pp. 768-779, 18 March, (2013)

23. Khandelwal Preeti, Sharma Ram Avatar, Agarwal Mala; *Journal of Plant Sciences* (2015).

24. M. S. Ambawade1, G.R. Pathade *International Journal of Science and Research* 6.14 (2013)

25. Masroor Qadri, Sarojini Johri, Bhahwal A Shah, Anamika Khajuria, Tabasum Sidiq, Surrinder K LattooMalik Z Abdin and Syed Riyaz-Ul-Hassan; *SpringerOpen Journal* 2:8 (2013)

26. Natalia Malfanova, Faina Kamilova, Shamil Validov,Andrey Shcherbakov, Vladimir Chebotar, Igor Tikhonovich and Ben Lugtenberg; *Microbial Biotechnology* (2011)

27. P. Chaudhry *Journal of Biodiversity and Ecological Sciences No.1, Issue4 Fall (*2011).

28. R E Jalgaonwala, B V Mohite and R T Mahajan *International Journal on Pharmaceutical and Biomedical Research (IJPBR)*Vol. 1(5), 136-141,(2010)

29. R.E. Jalgaonwala and R.T. Mahajan; *Journal of Agricultural Technology* Vol. 7(6): 1733-1741 (2011)

30. Ruby E Jalgaonwala; *International Journal Of Pharma And Bio Sciences* ;Volume 2/Issue1/ Jan March (2011)

31. Rupert AnandYumlembam And S.G. Borkar ; *Phytopath.* 67 (4) : 353-357 (2014)

32. SonaimuthuVisalakchi And JohnpaulMuthumary *International Journal Of Pharma And Bio Sciences* Vol.1/Issue-3/Jul-Sep.(2010)

33. Ulrich K., Ulrich A., and Ewald D; *FEMS Microbial Ecol*, 63:169-180, (2008).

34. Zareen Khan, Grant Guelich, Ha Phan, Regina Redman, and Sharon Doty; *International Scholarly Research NetworkISRN Agronomy*, Volume (2012)

3

Food Security in Rural Areas of Bihar: A Recent Perspective

Ranjeet Kumar Gupta and Shahnaz Jamil

University Department of Botany,
L. N. Mithila University, Darbhanga, Bihar-846004

INTRODUCTION

Food security is considered a pivotal factor for the sustainable development of communities and focus on this issue in rural areas. More specifically, it is of paramount importance in developing countries. Accordingly, this descriptive-analytical study aimed to evaluate the status of food security in rural areas of North Bihar. The term food security was first introduced at the World Food Summit in 1974. The meeting focused on the fact that food security is a mutual responsibility of all nations and its achievement requires international cooperation and serious measures (Jones *et al.,* 2013).

In fact, food and safety are necessary for political stability and creating peace among people (Helen *et al.,* 2002). Scientifically, food security is defined as a planned method designed to solve food and nutrition problems in the framework of codified programs and development management (Aziz *et al.,* 2020; Schleifer *et al.,* 2020).

In addition to the above-mentioned challenges, issues, such as improper methods of resource utilization, low level of knowledge in rural households, increasing decline in water resources' quality and quantity, increased inequality among rural families, environmental pollution, increased toxins, contamination of water and soil resources, and decreased soil fertility, have prevented sustainable food security. Meanwhile, maintaining and enhancing environmental, social, and economic grounds are crucial to achieve sustainable food security (Ahmadi Dehrashid, 2017).

Today, 13% of the population of developing countries are dealing with food insecurity (FAO, 2020; Pawlak *et al.,* 2020). In other words, about 98% of those who face malnutrition and food insecurity in the world are living in developing countries (Mota *et al.*, 2019; FAO, 2010). Since a considerable part of these people live in rural regions, improving the food security of rural households is of great importance (Babatunde *et al.*, 2007; Abdoellah *et al.*, 2020). Today, rural areas are facing food insecurity due to various reasons, including rapid growth in the population, low productivity in agriculture, lack of sustainable policies related to food security, macroeconomic instability, and sharp fluctuations in agricultural prices (Derrickson *et al.*, 2001; Anderman *et al.*, 2014).

Bihar is rich in natural resources, but food insecurity is still a significant concern. However, few studies still examine the relationship of socio dynamics in contributing to local ecosystems to create food security. In this study, we discussed how social dynamics contribute to normative structures, community habits and livelihoods to meet the living needs of rural communities in creating household food security and how they try to deal with worsening food insecurity through the local wisdom of rural communities.

Food security, with guarantees all human beings physical and economic access to the basic foods they need to lead active and healthy lives, bears multisectoral implications that go beyond the need for food production, supply and procurement. It is a dynamic process that is closely linked to the thematic topics of poverty, human capacity, creation of employment and generation of income in an ever changing spiral that can move upwards and downwards.

MATERIAL AND METHODS

Research Design

This study was an applied, qualitative, non-experimental, and descriptive-analytical research in terms of its orientation, the nature of the issue, extent and degree of control over variables, and methodology, respectively. The study was conducted using multi-criteria decision analysis methods and survey techniques.

Study Area

The research setting was five administrative villages of Kalyanpur block in the Samastipur district like Kudhwa, Belsandih, Singhia, Nawabganj and Janardanpur.

The instrument of data collection and survey

The panel members included researchers and specialists who had carried out extensive research on food security and its measurement and assessment. The meticulous assessment of the documents, the most related works were selected, and then, a number of researchers and authors, whose works were mostly related to the present research, were selected as the members of the specialist panel.

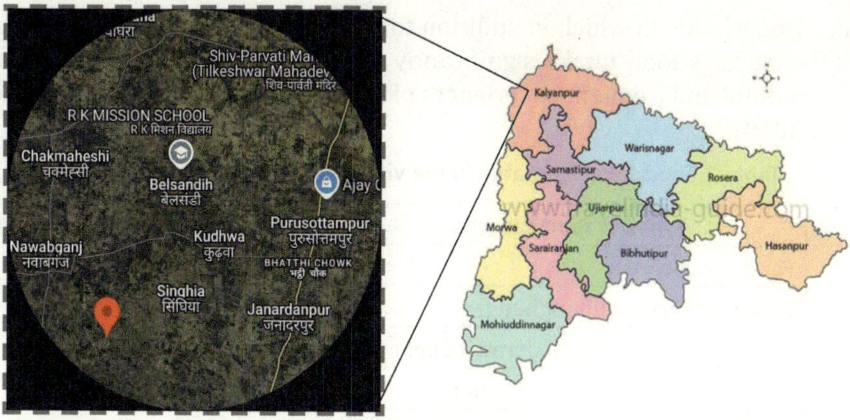

Fig-1. The site of the study area (location map of Samastipur district)

We also used the standard Department of agriculture questionnaire to evaluate food security within the area under study (Nord, 2009; Ramesh *et al.*, 2010). Notably, some of the indicators expressed in the questionnaire were localized based on the demographic characteristics of the rural community, and some other indicators were developed according to the knowledge of the target community. In addition, a researcher-made questionnaire was exploited to evaluate the consequences of food security. This instrument was able to measure the severity of food insecurity and family hunger. This instrument divided the studies' community into four groups/categories (with food security, with food insecurity without hunger, with food insecurity with moderate hunger, and with food insecurity with severe hunger) (Table 1).

Table 1 Typology of food security/insecurity in the villages of Samastipur district.

Types of food security	Code	Number of positive responses	
		Families with children under 18 years old (total score: 18)	Families without children under 18 years (Total score: 10)
Food security	0	0-2	0-2
Food insecurity without hunger	1	3-7	3-5
Food insecurity with moderate hunger	2	8-12	6-8
Food insecurity with severe hunger	3	13-18	9-10

a. Households that show little evidence of food insecurity or have not experienced food insecurity during the year

b. Households who are concerned about the inadequacy of food and manage the current situation by reducing the quality of food or reducing the amount of food consumption

c. Households in which food consumption is significantly reduced for adults, so that adults often experience hunger

d. Households in which in addition to reduction of adults' food intake, the children's food intake significantly affected, so that children also have painful and frequent experiences of hunger Source: (Nord 2009; Ramesh 2010)

Table 2 Food security status in the villages of Samastipur district

Types of food security	Frequency	Percent
Food security	12	20
Food insecurity without hunger	15	25
Food insecurity with moderate hunger	25	42
Food insecurity with severe hunger	8	13
Total	60	100

Table 3 Consequences of food insecurity in the villages of Samastipur district

Components	Variables	ID Code
Physical-skeletal	Agricultural land use changes	C1
	Horticultural land use changes	C2
	The weakening of the agricultural sector and the prosperity of service activities in the village	C3
Economic	Expansion of land and stock exchange transactions	C4
	Increases in land, housing and rental prices	C5
	Increasing the price of services	C6
	Increased inflation and rental income	C7
	Increasing income gap between rural and urban areas	C8
Social-cultural	Expansion of rural-urban migrations	C9
	Reduction of social capital in rural areas	C10
	Reduction of participatory and group activities in the village	C11
	Satisfaction with the quality of educational, social and cultural services	C12
	Satisfaction with the quality of health services	C13
	Satisfaction with the quality of the living environment and welfare services	C14
	Satisfaction with the quality of facilities and infrastructure	C15
	Satisfaction with the quality and quantity of services of rural financial institutions	C16

Components	Variables	ID Code
Environmental	Sustainability of water resources (improper operation, irrigation at inappropriate times)	C17
	Sustainability of soil and air resources (limited use of pesticides and chemical fertilizers, burning of crop residues)	C18
	Disruption of the natural ecosystem (burning of trees, hunting of birds, non-compliance with the number of livestock and pastures, pressure on natural resources and loss of ecological potential of resources)	C19
	Declining groundwater resources	C20
	Drying of rivers, springs and wells due to over-exploitation	C21
	Drilling unlicensed wells	C22
	Pressure on water resources and the occurrence of hazards (ground subsidence, landslides, erosion etc.)	C23

RESULTS AND DISCUSSION

According to the results, 80% of the villages suffered from food insecurity (25% in low food insecurity situation, 42% in moderate food insecurity situation and 13% in severe food insecurity situation). Evaluation of the factors affecting food insecurity demonstrated that economic, stability and availability were the most important components affecting food insecurity.

Evaluation of food security status showed that 20% of villagers in the studied area had food security, whereas 25, 42, and 13% had food insecurity without hunger, food insecurity with moderate hunger, and food insecurity with severe hunger, respectively (Table 2). Given the fact that a large part of the studies community had food insecurity, it is crucial to recognize the most important indicators and components that lead to the formation of food insecurity. According to the results the economic, stability and availability were the most important components affecting food insecurity.

According to the results of the study, food insecurity had consequences in the studied area, the most important of which are in the physical and environmental dimensions. In the former, indexes of C3 (weakening of the agricultural sector and prosperity of service activities in the village), C2 (horticultural land use changes), and C3 (agricultural land use changes) obtained the greatest weights. In the economic dimension, the indexes of C8 (increased income gap between rural and urban areas), C6 (increased cost of services), and C5 (increasing the cost of lands, houses, and rents) were the most important indicators, whereas, in the socio-cultural dimension, the indexes of C11 (reduction of participatory and group activities in the village), C9 (decreased social capital in villages), and C10 (expanding rural and urban migration) were the most significant (important) in dicators. In the environmental dimension, the indexes of C20 (declining groundwater resources), C22 (drilling unlicensed wells), C23 (pressure on water sources and occurrence of ground subsidence, landslides, and erosions), and C21 (drying of rivers, springs, and wells due to over-exploitation) had the most effect (Table 3).

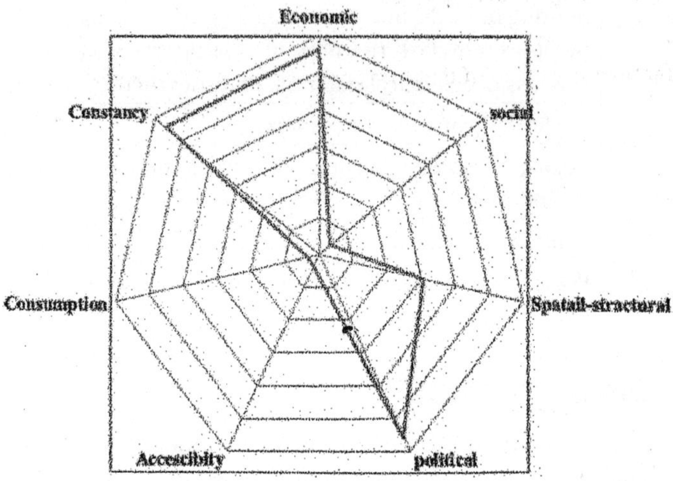

Fig. 2 Prioritizing the consequences of food insecurity in the study area

CONCLUSION

Since the low-income level of villagers, lack of job diversity and lack of access to adequate food are the main reasons for food insecurity, in this respect, it is suggested that decision-makers and policy-makers should pay more attention to the development of occupations and job diversity in these regions. According to the results of the present study, 20% of families had food security, whereas 25%, 42%, and 13% had food insecurity without hunger, food insecurity with moderate hunger, and food insecurity with severe hunger, respectively. Therefore, it could be concluded that most individuals had food insecurity. According to the results, changing the use of agricultural lands and the weakening of the agricultural sector in the studied area were the most important consequences of food insecurity in the villages of Kalyanpur block in the Samastipur district. The study was based on a cross-sectional survey of village administrators, and thus, results' analysis was fulfilled mainly by quantitative methods. Therefore, it is recommended that further studies be carried out using mixed methods (qualitative and quantitative) to yield more reliable and wider results. However, future researchers can fill this gap using some comprehensive meta-analysis-based studies.

REFERENCES

1. Derrickson JP, Sakai M, Anderson J. (2001) Interpretations of the "balanced meal" household food security indicator. *J Nutr Educ.* 33(3) : 155-60. https://doi.org/10.1016/S1499-40458(06)60185-1.

2. Helen H J. (2002). Food seurity and the food stamp program. *Am J Agric Econ.* 84(5):1215-8. https://doi.org/10.1111/1467-827600382.

3. Babatunde RO, Omotesho OA, Sholotan OS. (2007). Socio-economic characteristics and food security status of farming households in Kwata State, North-Central Nigeria. *Pakistan J Nutr.* 6(1) : 49-58.

4. Nord M. (2009). Food insecurity in households with children: prevalence, severity, and household characteristics. Economic information bulletin number 56. US Department of Agriculture.

5. Food and Agriculture Organization of the United Nations (FAO). (2010). The State of food and agriculture (SOFA). Women in agriculture closing the gender gap for development. Rome: Food and Agriculture Organization of the United Nations (FAO).

6. Ramesh T, Dotosty Motlagh AR, Abdollahi M. (2010). Prevalence of household food insecurity in the City of Shiraz and its association with socio-economic and demographic factors, 2008. *Iranian J Nutr Sci Food Technol.* 4(4):53-64.

7. Jones AD, Ngure FM, Pelto G, Young SL. (2013). What are we assessing when we measure food security? A compendium and review of current metrics. Adv Nutr. 4(5):481-505. https://doi.org/10.3945/an.113.04119.

8. Anderman TL, Remans R, Wood SA, DeRosa K, DeFries RS. (2014). Synergies and tradeoffs between cash crop production and food security: a case study in rural Ghana. Food security. 6(4):541-54. https://doi.org/10.1007/s12571-014-0360-6.

9. Mota AA, Lachore ST, Handiso YH. (2019). Assessment of food insecurity and its determinants in the rural households in Damot Gale Woreda, Wolaita zone, southern Ethiopia. *Agric Food Security.* 8(1):1-11. https://doi.org/10.1186/s40066-019- 0254-0.

10. Abdoellah OS, Schneider M, Nugraha LM, Suparman, Y, Voletta CT, Withaningsih S, Hakim L. (2020). Homegarden commercialization: extent, household characteristics, and effect on food security and food sovereignty in Rural Indonesia. *Sustain Sci.* 15(3):797-815. https://doi.org/10.1007/s11625-020-00788-9.

11. Aziz N, Nisar QA, Koondhar MA, Meo MS, Rong K. (2020). Analyzing the women's empowerment and food security nexus in rural areas of Azad Jammu & Kashmir, Pakistan: By giving consideration to sense of land entitlement and infrastructural facilities. Land Use Policy. 94:104529. https://doi.org/10.1016/ j.landusepol.104529.

12. Food and agriculture organization of the United Nations (FAO). Our World in Data. 2020. https://ourworldindata.org/hunger-andundernourishment#whatshare-of-people- are-undernourished.
13. Pawlak K, Kolodziejczak M. (2020). The role of agriculture in ensuring food security in developing countries: considerations in the context of the problem of sustainable food production. Sustainability. 12(13):5488. https://doi.org/10.3390/su12135488.
14. Schleifer P, Sun Y. (2020). Reviewing the impact of sustainability certification on food security in developing countries. *Global Food Security.* 24:100337. https://doi.org/10.1016/j.gfs.2019.100337.

4

Mycotoxins Requisite Levels Required for Better Consumption of Food Grains as a Food Security in India

B. L. Gupta[1] and Gajendra Prasad[2]

[1]*Dept. of Zoology, J.K. College, Biraul, Darbhanga, Bihar-846004, India*
[2]*University Department of Botany, L.N. Mithila University, Darbhanga-846004, Bihar*

INTRODUCTION

Food grains and other field crops are generally found to contain unsafe biochemicals like mycotoxins along with other food contaminations. When these substances are consumed at certain levels for a longer time then health risks are prone to humans as well as animals. Consuming grains or other foods contaminated with certain mycotoxins can be fatal if the toxins are present at very high levels. Long-term exposure to mycotoxins can increase cancer risk, liver diseases, perforation in digestive tract, the immune system suppression, etc. Such mycotoxins infected food stuff has to be eliminated from food chain that leads to loss to farmers and the governments. More than $1 billion annual loss in 2017 USD from mycotoxins has been recorded in US crops (Gitz, V. et al 2016).

Toxicogenic fungi produce mycotoxins that grow on a variety of crops and foodstuff and can penetrate deep into food and do not just grow on the surface. Such fungal contamination in food is almost unavoidable because it may occur in any stage of the supply chain, from the field to eating food (Kos J et al 2023; Perrone, G et al 2020). The main producers of mycotoxins are species that belong to the genera *Aspergillus, Alternaria, Fusarium* and *Penicillium*. The most prevalent mycotoxin families are aflatoxins, fumonisins, ochratoxin A, patulin, tenuazonic acid, trichothecenes and zearalenone (El Darra, N et al 2023; Garcia, D et al 2009; Luo, Y et al 2014)

The food security has to come into existence because of having standard food quality and a good health. The policies on food security have been embarked by different countries, some with loose inaction and some with stringent policies and fundamentals. The risk of global food insecurity has been increased by many factors, such as inadequate economic distribution, low availability of commodities and climate changes. The climate is an important issue to ensure food security because it affects the quality and quantity of food production and infections by toxicogenic fungi (Gitz, V. et al 2016).

MYCOTOXINS GENERATIONS

Mycotoxins are an important food safety concern for grains and other field crops. Mycotoxins are toxic byproducts of mold infestations affecting as much as one-quarter of global food and feed crop output. Food contaminated with mycotoxins, particularly with aflatoxins, can cause sometimes-fatal acute illness, and is associated with increased cancer risk.

Mycotoxin contamination is recognized as an unavoidable risk. Many factors that influence the level of contamination in cereals and grains are environmental, such as weather and insect infestation and therefore are difficult or impossible to control the growth of molds with 100% but can be minimized so that the food may be consumable for both the animals and humans.

Mycotoxins are produced by certain fungi (e.g., Aspergillus ssp., Penicillium ssp., and Fusarium ssp.) that grow on human food and animal feed ingredients such as corn, sorghum, wheat, barley, peanuts, and other legumes and oilseeds. Five broad groups of mycotoxins - aflatoxin, vomitoxin, ochratoxin A, fumonisin, and zearalenone, are commonly found in food and feed grains. Among mycotoxins, probably the most widely recognized risk comes from aflatoxins. Aflatoxins are extremely potent carcinogenic and mutagenic substances that first came into the public spotlight and were formally identified in the early 1960s following the deaths of more than 100,000 young turkeys on a poultry farm in England.

The fungi that produce mycotoxins can emerge either in the field like in soil, decaying vegetation, and grains undergoing microbiological deterioration or during postharvest transportation or storage. Temperature stress is an important cause of fungi growth on crops in the field, and high moisture content and temperature are associated with the growth of fungi in stored grain. Detection and control of the fungi is a continuous concern since the fungi may become established and remain with the commodity anywhere along the production, storage, transportation, and processing chain. A further concern is that the absence of visible mold does not guarantee the grain is free from mycotoxin, and cooking or processing the food product does not necessarily rid it of mycotoxin contamination. For example, molds contaminated with aflatoxins have been isolated in processed food products such as bread, macaroni, cooked meat, cheese, and flour (Guerzoni, M.1999).

MYCOTOXINS AND DISEASES

Food safety concern is potential exposure to mycotoxins through consumption of food from contaminated crops, which can produce acute or long-term health problems. Consuming food products that contain high levels of certain mycotoxins can cause the rapid onset of mycotoxicosis, a severe illness characterized by vomiting, abdominal pain, pulmonary edema, convulsions, coma, and in rare cases, death. Although lethal cases are uncommon, acute illnesses from mycotoxins, particularly aflatoxins (aflatoxicosis), have been reported from many parts of the world, usually in developing countries (Miller, J.D. 1996).

The food safety concern becomes the key regulation than acute illness in the long-term effects of lower-level mycotoxin consumption, particularly the risks of cancer and immune deficiency. Aflatoxin B1 has been mentioned on the list of known human carcinogens by the International Agency for Research on Cancer (IARC) in 1988, and other mycotoxins are suspected or known to be carcinogenic or to have other adverse health consequences. Aflatoxins are a particular concern for populations with a high incidence of hepatitis B because the relative rate of liver cancer in people with hepatitis B is up to 60 times greater than normal when those people are exposed to aflatoxin (Miller, 1996). In addition to direct risks to humans from consumption of mycotoxin-contaminated grains, there are indirect health risks to those who consume animal products containing residues of carcinogenic mycotoxins. Mycotoxins can be detected in meat, milk, and eggs from animals that have consumed feed ingredients containing mycotoxins. Many countries have tolerance standards for mycotoxin residues in these products (Miller, J.D. 1995). Another concern related to the consumption of mycotoxin-contaminated feed by livestock is the potential for economic losses from animal health and productivity problems. Aflatoxins in feed are known to be associated with liver damage in animals, reduced milk and egg production, poor weight gain, and recurrent infections due to immunity suppression. The young of any particular species are most vulnerable, but the degree of susceptibility varies by species.

AFLATOXINS IN FOOD PRODUCTS

Aflatoxins which were studied in maize seeds in our laboratory showed a comprehensive decline in many glycolytic enzymes and metabolites showing a degradation of maize seeds ingredients along with aflatoxins present in the maize seeds (Gajendra, P and Gupta, B.L. 2019). The optimal concentration of aflatoxins as referred by international bodies particularly USA is 0.5 to 20 μg/kg of seeds, may be recommended for commercial consumption of maize seeds by humans and animals also as it is understood that mycotoxins can not be eliminated from food chain. Some countries may set new standards on certain mycotoxins (Batista, L. R., & O'Brien, S. 2021). In India it requires a compulsory regulator for mycotoxins present in food grains and cereals for our food safety.

Aflatoxins play a significant role in food security due to their toxic effects on food safety, agricultural productivity, and public health.

1. **Health Risks**: Aflatoxins are highly toxic compounds produced by certain molds, especially *Aspergillus* species that can contaminate crops

such as peanuts, corn, and grains. Their presence poses serious health risks, including liver damage, immune suppression, and increased risk of cancer. Food safety may be guaranteed by controlling aflatoxin levels protecting public health (FAO 2020).

2. **Economic Loss**: High levels of aflatoxins can lead to a significant economic loss for farmers and the food industry. Contaminated crops may be rejected in the market, leading to reduced income for farmers and increased food prices.
3. **Agricultural Inabilities**: Aflatoxin contamination leads to poor agricultural practices, such as inadequate drying, poor storage conditions, and pest management. The use of modern technology may limit contamination assisting to food security.
4. **Regulatory policies**: Enforcing regulations on aflatoxin levels in food products is crucial for ensuring food safety. Governments and organizations enact and implement effective policies in monitoring and controlling aflatoxins levels to protect consumers (Peraica, M. et al 1999).
5. **Public Awareness**: There must be awareness about aflatoxins, their sources, and prevention strategies among the farmers and consumers to make food safety and security. They must also know on proper storage and handling practices (WHO, 2021).
6. **Researches and Implementation**: Researches on aflatoxins in crops by biotechnological means on the development of resistant crop varieties, biocontrol methods, and improved post-harvest practices, can secure food security (Kumar, P., et al. 2020).

MYCOTOXINS AND CONSUMABLE LEVELS

Mycotoxin levels in food products are regulated to ensure safety for human consumption. Different mycotoxins have different permissible limits, which can vary by country and food type. Here are some commonly regulated mycotoxins and their typical limits:

1. Aflatoxins:

- Commonly found in grains, nuts, and seeds.
- Maximum allowable levels can be as low as 4 to 20 micrograms per kilogram (μg/kg) in various foods, depending on the type and country.

2. Ochratoxin A:

- Found in cereals, coffee, and dried fruits.
- Permissible limits may range from 2 to 5 μg/kg.

3. Fumonisins:

- Primarily associated with maize (corn) products.
- Limits are often set at around 1,000 to 4,000 μg/kg for different food products.

4. Zearalenone:

Often found in cereals and grains.

Limits can range from 20 to 1,000 μg/kg, depending on the specific product.

5. Deoxynivalenol (DON):

- Commonly present in wheat and barley.
- Acceptable levels are often set at 1,000 to 2,000 μg/kg.

Mycotoxins production are mostly affected by environmental factors in India which has a tropical climate, which is suitable for the growth of moulds. Due to a different seasons in India the mycotoxins productions seems to vary from one state to others and therefore, their testing may lead to variable amounts of mycotoxins for a similar product. Handling and storage conditions are important risk factors associated with post-harvest mycotoxin accumulation. The extent of AF production in cereals depends on temperature, moisture, soil type and storage conditions. Among cereals, rice is mostly contaminated with AFs in the natural environment. While nuts are highly susceptible to AF contamination both in the field and during storage, spices are significantly affected by storage and processing conditions (Mahato D.K. et al 2019). For this, the Indian food industry has implemented several internal monitoring mechanisms for domestic control.

Mycotoxin surveillance in food and feed is conducted on a regular basis by the country's nodal regulatory agencies as part of pre-export testing. These control methods are mostly preventive in nature, including the implementation of GAPs at the field level and adequate crop drying after harvest. The International Crops Research Institute for the Semi-Arid Tropics (ICRISAT) in Hyderabad, for instance, is conducting extensive research into methods to prevent crop contamination before harvest. Many efforts simply comprise removing mycotoxin-contaminated goods from the food supply chain via governmental screening and regulatory programmes. To reduce mycotoxin contamination, governmental agencies have implemented advanced agricultural technologies such as GAPs, good manufacturing practice (GMP) and good storage practice (GSP). The crop-specific institutes under ICAR and the state agricultural universities have published numerous GAP-based pre-harvest guidelines (e.g., fertiliser and pesticide management, weed management, water and nutrient management, among others). Numerous post-harvest techniques (e.g., drying, maintaining proper moisture levels, sorting damaged or infested grains and destroying contaminated crops) are adopted, too. Additionally, biological control has been proposed as a possible method for limiting the growth of mycotoxin-producing fungal communities (FASSAI, 2018). However, strict adherence to these practices is still required.

FSSAI: A REGULATORY BODY IN INDIA

The FSSAI is the chief regulatory body responsible for building an effective food safety network to provide scientific information to consumers regarding safe and healthy food. On 20 October 2020, this agency published a revised manual on mycotoxin analysis (Manual Report, 2020), where MLs of all concerned toxins in food matrices were compiled (Gazette Report,2022). These include Afs in peanuts (Dhanshetty M. et al 2021), deoxynivalenol (DON) in wheat (Dhanshetty M. P. et al 2021), ZEN (Mishra S. et al 2013) and FUM (Prabhakar C. and Ghadevaru S., 2020) in maize, and PAT in apple juice (Kamle M. et al 2019). It made the manual available to the public to facilitate foreign trade.

In accordance with Section 43 of the Food Safety and Standards (FSS) Act of 2006, the agency has granted approval to several food testing facilities that are accredited to the ISO/IEC17025:2017 standard. Mycotoxin levels in local and imported food items are continuously monitored in three national food laboratories (NFL), one of which is located in Ghaziabad and the others in Kolkata and Mumbai (Shinde R., et al 2021; FSSAI Report 2021). Under the FSS Act of 2006, these apex food laboratories are responsible for food quality testing. The NFLs in Ghaziabad and Mumbai are operated through public–private partnerships. There are 227 primary food laboratories accredited by the National Accreditation Board for Testing and Calibration Laboratories (NABL) (FSSAI Report 2022), which are meant to cover all zones (Northern, Southern, Eastern and Western) of the country.

Furthermore, FSSAI has recognised 19 referral laboratories (RLs) for analysing appeal samples that are rejected on account of above-ML residue issues (List of Referral Laboratories, 2022). In addition, 11 national reference laboratories (Order NRL Approval, 2022) have been recognised for establishing nationwide standards for routine analysis, developing new and reliable methods, validating such testing methods and organising proficiency testing (PT). These laboratories are part of the most prestigious research and development organisations in the nation and are located in Mysore (1), Mohali (1), Kochi (3), Pune (1), Anand (1), Lucknow (1), Hyderabad (1), Kolkata (1) and Gurugram (1). Additionally, two ancillary laboratories in Chennai and Kolkata have been identified to assist the NFLs in organising PTs. Whereas, National Reference Laboratories, located at NRL-NRCG and Trilogy Analytical Laboratory Pvt. Ltd., Hyderabad, have been assigned for mycotoxin analysis and providing reference materials, respectively. Both of these laboratories offer support in analysing imported samples, particularly when a primary food laboratory rejects them for having mycotoxin contamination levels exceeding the national ML.

Concurrently, the FSSAI's Scientific Panel on "Methods of Sampling and Analysis" makes insightful recommendations on existing testing methods and incorporates new test methods for analysing new parameters (commodity-ML combinations). The agency also provides courses on the microbiological and chemical aspects of mycotoxin testing to laboratory personnel from FSSAI-notified and central food laboratories, among others (FSSAI Annual Report 2019-20). In 2019, two such "Training of Trainers" programmes on mycotoxin analysis took place at NRL-NRCG and the Export Inspection Agency (EIA), Kochi (FSSAI Annual Report 2019-20). Additionally, the FSSAI provides portable lateral flow assay readers to all regulatory laboratories in various states and union territories to facilitate mycotoxin monitoring. This action was taken while keeping in mind that biological toxins can cause permanent health damage.

Under Section 16 (2) (f) of the FSS Act, 2006, the agency has published specific manuals on analytical methods for different food commodities with a "one parameter–one method" approach. The first mycotoxin analysis manual, released in 2016, was written with the primary goal of providing thorough and up-to-date methodologies for regulatory compliance. After several revisions, the latest version was published in 2021, in which several validated methods (mostly contributed by NRL-NRCG) of AFs, OTA and PAT were incorporated. This analytical manual summarises regulatory limits (Table 1), safety requirements and 22 well-defined analytical methods comprising sample preparation as well as instrumental analysis summarised in Table 2. By making these records accessible to the public, the FSSAI has significantly improved current mycotoxin testing practices.

Table 1: Current FSSAI regulatory limits for mycotoxins in India.

Mycotoxin	Food Product	FSSA(I) Regulatory Limit (μg /Kg)
Afs	Cereal and Cereal Products	15
	Pulses	15
	Nuts	
	Nuts for further processing	15
	Ready to eat	10
	Dried figs	10
	Oilseeds or oil	
	Oilseeds for further processing	15
	Ready to eat	10
	Spices	30
	Betelnut/Arecanut	15
AF M1	Milk	0.5
OTA	Wheat, barley and rye	20
PAT	Apple juice and Apple juice ingredients in other beverages	50
DON	Wheat	1000

Table 2: Summary of currently practised mycotoxin detection methodologies as per FSSAI Manual of Methods of Analysis of Foods—Mycotoxins.

Mycotoxin	Matrix	Detection Method	LOD	LOQ	Linearity Range
Total AFs	Groundnuts and groundnut products, oilseeds and food grains	Thin-Layer Chromatography (TLC)-Ultraviolet (UV)	-	-	-
Total AFs	Food and feed	TLC-UV: Romer Mini Column Method	-	Almonds ≥ 5 ng/g, corn, cotton seed meals, nuts ≥ 10 ng/g, mixed feeds ≥ 5 ng/g	-
Total AFs	Corn and groundnuts	TLC-UV	-	B1, G1 = 5 to 50 ng/g, B2, G2 = 3 to 15 ng/g	-
Total AFs	Corn and peanut powder/ butter	HPLC-FLD (Derivatised)	0.3 ng/g	-	B1, G1 = 0.25 to 2 ng/ mL, B2, G2 = 0.125 to 1 ng/ mL
AFs (B1, B2 and G1)	Corn, cottonseed, peanut, peanut butter	ELISA (Immuno-dot screen cup)	-	Cottonseed, peanut ≥ 20 ng/g, corn, raw peanuts ≥ 30 ng/g	-
AFs (B1, B2 and G1)	Corn	ELISA (Afla-20 cup test)	-	≥20 ng/g	-
AF B1 and Total AFs	Peanut butter, pistachio paste, chilli, paprika powder and other spice powders, dried figs and other dried fruits	HPLC-FLD (IAC cleanup and derivatised)	-	-	B1, G1 = 0.4 to 3.6 ng/ mL, B2, G2= 0.08 to 0.72 ng/mL

AF B1	Baby food	HPLC-FLD (IAC cleanup and derivatised)	-	≥0.1 ng/g	0.01 to 0.07 ng/mL
Total AFs + their derivatives (G2a, B2a)	Olive oil, peanut oil and sesame oil	HPLC-FLD (IAC cleanup and derivatised)	-	2 to 20 μg/kg	B1 = 0.4 to 10 ng/mL, G1 = 0.2 to 5 ng/mL, B2, G2 = 0.1 to 2.5 ng/mL
Total AFs	Peanut, peanut products and cereals	UHPLC-FLD (Direct method)	-	0.008 μg/kg for B1 and G1, 0.003 μg/kg for B2 and G2	0.02 to 10 ng/g
AFs M1 and M2	Milk	LC-FLD	-	-	-
AF M1	Milk	LC-UV Vis	-	>0.02 ng/mL	0.05 to 1 ng/mL
Total AFs	Matrices other than described above	HPTLC-UV	-	-	-
DON	Wheat	TLC-UV	-	≥300 ng/g	-
OTA	Barley	TLC-UV	-	-	-
OTA	Barley	HPLC-FLD(IAC Cleanup)	0.1 ng/g	>1 ng/g	0.5 to 10 ng/mL
AF + OTA	Cereals	UHPLC-FLD	AF = 0.02 ng/g, OTA = 0.1 ng/g	-	0.02 to 10 ng/mL for AF, 0.1 to 10 ng/mL

CONCLUSION

Mycotoxins are real threat to humans as well as animals because they are toxic in natures which have been considered as mother of generation of most of diseases. If they are present in low amount then mycotoxins are not fatal and suitable for consumption. Due to verities of seasons in India, the growth of moulds on crops has been found different. Therefore, the mycotoxins amount for a single crop becomes different as their handling and storage was found to be different. The government of India, however, incorporated Food Safety and Standards Act of 2006 but we need laboratories all around the country to facilitate for testing. We do not know mycotoxins present in our daily food like bread, butter, milk, cereals, etc and the wrappers in which food has been packeted, does not describe the levels of mycotoxins present in the food. We do not know the diseases which come out after consuming the mycotoxins infested food items. It will be a real research and innovation to find the real facts in controlling diseases in India. We also do not know that how much mycotoxins present in street foods in India. We are going through many challenges and the government policies must be formulated in this regard to open door for food safety and security in near future in every state of India.

REFERENCES

1. Batista, L. R., and O'Brien, S. (2021). Mycotoxins: An Overview of Their Impact on Food Security. *Food Control*, 125, 107877.
2. Dhanshetty M., Elliott C.T., and Banerjee K.(2021) Decontamination of aflatoxin B1 in peanuts using various cooking methods. *J. Food Sci. Technol.* 2021;58:2547–2554. doi: 10.1007/s13197-020-04761-3.
3. Dhanshetty M.P., and Malib G.V. (2021) Isolation, Characterisation of Aflatoxin Producing Fungi from Contaminated Peanut and Rice, their Fate on Aflatoxin Production. YMER. ;20:565–572.
4. El Darra, N.; Grimi, N. and Watson, I.A.; (2023) El Khoury, A. Editorial: Risk assessment of mycotoxins in food. *Front. Nutr.* , *10*, 1145998.
5. Food and Agriculture Organization (FAO). (2020). Mycotoxins in Food: Methods of Sampling and Analysis.
6. FSSAI Publishes Guidance Note of Aflatoxins on 18 July (2018). (accessed on 17 November 2022).
7. Gajendra, P. and Gupta, B.L. (2019). Mycotoxins Contamination and strategies in Plant Disease Management: A Review. In "Recent Trends in Plant Disease Management" (Eds. Sheela et. al.,) Today & Tomorrow's Printers & Publishers, New Delhi. pp-333-347.
8. Garcia, D.; Ramos, A.J.; Sanchis, V.; Marín, S. (2009) Predicting mycotoxins in foods: A review. *Food Microbiol.* , *26*, 757–769.

9. Gazette Notification: Crop Contaminants and Naturally Occurring Toxic Substances. (accessed on 8 April (2022)

10. Gitz, V.; Meybeck, A.; Lipper, L. Young, C.D.; Braatz, S. (2016). Climate change and food security: Risks and responses. *Food Agric. Organ. United Nations (FAO) Rep. 110*, 3–36.

11. Guerzoni, Maria. (1999). "Position Paper on Some Aspects Concerning Foodborne Diseases and Food Toxicity in the Mediterranean Areas." The Techno-Economic Analysis Network for the Mediterranean (TEAM), Working Group on Food Technology and Toxicity in the Mediterranean.

12. Kamle M., Mahato D.K., Devi S., Lee K.E., Kang S.G., Kumar P. (2019). Fumonisins: Impact on agriculture, food, and human health and their management strategies. *Toxins.* 11:328. doi: 10.3390/toxins11060328.

13. Kos, J.; Anić, M.; Radić, B.; Zadravec, M.; Janić Hajnal, E.; Pleadin, J. (2023) Climate change—A global threat resulting in increasing mycotoxin occurrence. *Foods 12*, 2704.

14. Kumar, P., *et al.* (2020). Mycotoxin Contamination of Food: Current Status and Future Perspectives. *Critical Reviews in Food Science and Nutrition*, 60 (17), 2924-2944.

5

Impact of Medicinal Plant Extracts on Amino Acid Content of Aflatoxin-Treated Maize Seeds (*Zea mays* L.)

Ragni Kumari and Gajendra Prasad

University Department of Botany,
L.N. Mithila University, Darbhanga-846004, Bihar

INTRODUCTION

In Bihar state, maize crop is cultivated in all the three important seasons, like summer, monsoon and winter that correspond to specific climatic or agronomic conditions. The harvested maize crop is kept under storage for optimum temperature for future consumption for a longer period due to a slow degradation process. The development of fungi infestation in post-harvesting maize is favored by several factors and co-factors like availability of inoculum, moisture, aeration, the physical integrity of seeds, optimum temperature and nature of the compositions. Maize cultivation is a low-cost input system, farmers tend to increase planting area rather than intensify production.

Particularly in Bihar, natural contamination of mycotoxins in agricultural commodities is fairly high because of its socially and economically backwardness, unhygienic, faulty and outdated agricultural practices and storage methods, moderate to high temperature as well as excessive humidity. Under favorable conditions, the maize grains become an excellent platform for the growth of mycotoxin-producing fungi. Climatic conditions and environmental factors particularly optimum temperature of 30-35^{0}C (Abdel- Hadi *et. al.*, 2012) and water activity, play a vital role in determining fungal species influencing their growth and mycotoxins production (Magan and Medina, 2016).

Aflatoxins are too common contaminants of crucial agricultural commodities including maize and peanuts. *Aspergillus flavus* and related species produce aflatoxins, which are secondary metabolites that can adversely affect human health and food security in warm agricultural area (Cotty *et. al.*, 1994; Bandyopadhyay *et. al.*, 2016). Aflatoxin B_1, which is classified as a group 1a carcinogen by the International Agency for Research on Cancer (IARC, 2002), is the most common and toxic of the four major aflatoxins B_1, B_2, G_1 and G_2.

Aflatoxin can become a serious case, when the moist grain cannot be arid quickly. Aflatoxin content has become a key factor affecting the exportation of maize grain.

In Bihar, only *Aspergillus flavus* and *Aspergillus parasiticus* have been found to be aflatoxin producing fungi.

BITTER PLANTS

The one with bitter leaf extract has a lower microbial count because bitter leaf extract lowered the bacterial and fungal growth due to it serves as an antimicrobial agent that inhibits microbial growth (Okafor, 2007; Munaya, 2013; Wang S., 2006). Bitter plants like *Andrographis paniculata, Azadirachta indica* and *Adhatoda vasica* have antifungal activity (Bhonde *et. al.,* 1999; Thanaboripat *et. al.,* 2000; Ramchandran and Sankaranarayanan, 2013; Sharma and Sanadhya, 2017; Singh *et. al.*, 2017).

Few reports have been available, on the impact of bitter plant extracts against microorganisms especially *A. flavus* and production of aflatoxin in SMKY liquid medium as well as PDA Medium (Bilgrami, KS. *et. al.,* 1980; Ranjan KS. *et. al.,* 1991; Masood and Ranjan, 1991; Ansari and Shrivastava, 1991; Sinha *et. al.,* 1993; Masood *et. al.,* 1994; Prasad *et. al.,* 1994; Thanaboripat *et. al.,* 2000; Choudary, 2002; Sidhu, O.P. *et. al.,* 2009; Thanaboripat, D. 2011; Zakaria *et. al.,* 2014; Ogo-Oluwa and Liamngee Kator, 2016; Sinha, B.K. 2018).

Powders of various spices, herbs and essential oils have been also reported to have antimicrobial activity against aflatoxin-producing fungi and some of them also inhibit aflatoxin production (Thanaboripat *et. al.,* 1989; Mahmood, 1994; Thanaboripat *et. al.,* 1997; Thanaboripat, 2003; Thanaboripat *et. al.,* 2004; Krishnamurthy and Shashikala, 2006; Razzaghi-Abyaneh *et. al.,* 2008; Reddy *et. al.,* 2010).

MATERIAL AND METHODS

Collection of Materials

Substrate

Maize seeds (*Zea mays L.*) of Syngenta Sugar 75 Sweet corn obtained from Pooja seeds, Agricultural seed store, Bazar Samiti Road, Shivdhara, Darbhanga, Bihar, 846004, were used throughout the experiments.

Toxins

Standards of aflatoxin B_1 was available in univ. Dept. of Botany, L. N Mithila University, Darbhanga.

Bitter Plants (*Andrographis paniculata, Azadirachta indica* and *Adhatoda vasica*) were obtained from the Janta Nursery, Karamganj, west of Naka- 06, Darbhanga.

Effect on Amino Acid (Essential and non- essential)

2 g powders of control and treated seeds were crushed with 100 ml of 80% ethanol separately in glass homogenizer and were boiled on a water bath for 20-30 minutes. Subsequently these were cooled and filtered. The filtrate was evaporated to dryness on a water bath and to the residue 2 ml of 70% ethanol was added. The solution thus obtained was centrifuged at 3000 r.p.m. for 20 minutes and the supernatant was stored to analyse the amino acids qualitative and quantitatively also.

Preparation of TLC plate

Thin layer chromatographic technique suggested by **Stahl, (1965)** was adopted for the detection of amino acids. A plate with 0.25 mm thick layer of silica gel- G was prepared by mixing the silica gel- G and distilled water (6:12 w/v).

Amino acid

0.01 ml of each alcoholic extract was loaded on the TLC-plate along with the standard solutions of amino acids and was developed in n-butanol: acetic acid: water (4:1:1, v/v) solvent system. After development, the plate was air dried and subsequently sprayed with 0.3% ninhydrin reagent (300 mg ninhydrin dissolved in 100 ml of n-butanol + 3 ml glacial acetic acid). After half an hour the chromatoplates were placed in oven at 100°C for 1-2 minutes. The spots developed were identified by comparing the R.f values of the unknown compounds with that of known amino acids loaded on the same plate. Quantitative estimation of amino acid was measured by the help of dilution technique.

RESULTS AND DISCUSSION

Amino acid

AFT-B_1 individually in various concentration up to 2 ppm exhibited marked inhibition in amino acids contents of germinating maize seeds depicted in Table-01.

Table- 01: Efficacy of Aflatoxin B1 on Amino acid contents of germinating maize seeds

Conc. Of AFT-B_1 (µg/l) (ppm)	L-Leucine		L-Methionine		DL (+) Alanine		Glutamic acid		Aspartic Acid	
	Amount (S.E)	Difference with Control (% inhibition)	Amount (S.E)	Difference with Control (% inhibition)	Amount (S.E)	Difference with Control (% inhibition)	Amount (S.E)	Difference with Control (% inhibition)	Amount (S.E)	Difference with Control (% inhibition)
00	1.06 (0.05)	-	0.15 (0.03)	-	0.62 (0.03)	-	1.53 (0.07)	-	0.49 (0.04)	-
100	0.82 (0.03)	0.24 (22.64)	0.12 (0.02)	0.03 (20.00)	0.43 (0.03)	0.19 (30.64)	0.87 (0.04)	0.66 (43.13)	0.42 (0.04)	0.07 (14.28)
250	0.58 (0.04)	0.48 (45.28)	0.08 (0.02)	0.07 (46.67)	0.31 (0.03)	0.31 (50.00)	0.61 (0.03)	0.92 (60.13)	0.36 (0.03)	0.13 (26.53)
500	0.43 (0.03)	0.63 (59.43)	0.08 (0.01)	0.07 (46.67)	0.14 (0.02)	0.48 (77.42)	0.43 (0.03)	1.10 (71.89)	0.22 (0.02)	0.27 (55.10)
1000	0.25 (0.03)	0.81 (76.41)	0.05 (0.01)	0.10 (66.67)	-	-	0.38 (0.03)	1.15 (75.16)	0.09 (0.01)	0.40 (81.63)
2000	0.22 (0.01)	0.88 (79.24)	-	-	-	-	0.34 (0.03)	1.19 (77.78)	-	-

Table- 02: Efficacy of *Andrographis paniculata* (A) on Amino acid contents of germinating maize seeds

Conc. Of Plant A (µg/l)	L-Leucine		L-Methionine		DL (+) Alanine		Glutamic acid		Aspartic Acid	
	Amount (S.E)	Difference with Control (% inhibition)	Amount (S.E)	Difference with Control (% inhibition)	Amount (S.E)	Difference with Control (% inhibition)	Amount (S.E)	Difference with Control (% inhibition)	Amount (S.E)	Difference with Control (% inhibition)
00	1.06 (0.005)	-	0.12 (0.001)	-	0.61 (0.005)	-	1.58 (0.023)	-	0.46 (0.005)	-
100	1.04 (0.023)	0.02 (1.88)	0.16 (0.005)	0.04 (-33.3)	0.67 (0.011)	0.06 (-9.83)	1.06 (0.023)	0.52 (32.91)	0.45 (0.011)	0.01 (2.17)
250	1.04 (0.011)	0.08 (1.88)	0.14 (0.005)	0.02 (-16.6)	0.69 (0.011)	0.08 (-13.11)	0.86 (0.011)	0.72 (45.56)	0.34 (0.024)	0.12 (26.08)
500	1.03 (0.005)	0.03 (2.83)	0.13 (0.005)	0.01 (-8.3)	0.63 (0.017)	0.02 (-3.27)	0.75 (0.005)	0.83 (52.53)	0.27 (0.024)	0.19 (41.30)
1000	1.05 (0.005)	0.01 (0.94)	0.14 (0.005)	0.02 (-16.6)	0.62 (0.005)	0.01 (-1.63)	0.65 (0.011)	0.93 (58.86)	0.17 (0.011)	0.29 (63.04)
2000	1.05 (0.017)	0.01 (0.94)	0.17 (0.011)	0.05 (-41.6)	0.60 (0.017)	0.01 (1.63)	0.32 (0.011)	1.26 (79.74)	-	-

Table- 03: Efficacy of *Azadirachta indica* (B) on Amino acid contents of germinating maize seeds

Conc. Of Plant B (µg/l)	L-Leucine		L-Methionine		DL (+) Alanine		Glutamic acid		Aspartic Acid	
	Amount (S.E)	Difference with Control (% inhibition)	Amount (S.E)	Difference with Control (% inhibition)	Amount (S.E)	Difference with Control (% inhibition)	Amount (S.E)	Difference with Control (% inhibition)	Amount (S.E)	Difference with Control (% inhibition)
00	1.05 (0.005)	-	0.12 (0.25)	-	0.63 (0.06)	-	1.53 (0.14)	-	0.47 (0.11)	-
100	1.04 (0.008)	0.01 (0.95)	0.15 (0.10)	0.03 (-25)	0.65 (0.10)	0.02 (-3.17)	1.08 (0.14)	0.45 (29.41)	0.46 (0.13)	0.01 (2.12)
250	1.04 (0.005)	0.01 (0.95)	0.12 (0.09)	0.00 (0.00)	0.68 (0.18)	0.05 (-7.93)	0.86 (0.10)	0.67 (43.79)	0.42 (0.16)	0.05 (10.63)
500	1.06 (0.005)	0.01 (-0.95)	0.11 (0.22)	0.01 (8.33)	0.66 (0.14)	0.03 (-4.76)	0.73 (0.11)	0.8 (52.28)	0.26 (0.21)	0.21 (44.68)
1000	1.07 (0.001)	0.02 (-1.90)	0.13 (0.11)	0.01 (-8.33)	0.65 (0.12)	0.02 (-3.17)	0.62 (0.25)	0.91 (59.47)	0.18 (0.23)	0.29 (61.70)
2000	1.04 (0.005)	0.01 (0.95)	0.14 (0.08)	0.02 (-16.66)	0.61 (0.28)	0.02 (3.17)	0.33 (0.13)	1.2 (78.43)	-	-

Table- 04: Efficacy of *Adhatoda vasica* (C) on Amino acid contents of germinating maize seeds

Conc. Of Plant C (µg/l)	L-Leucine		L-Methionine		DL (+) Alanine		Glutamic acid		Aspartic Acid	
	Amount (S.E)	Difference with Control (% inhibition)	Amount (S.E)	Difference with Control (% inhibition)	Amount (S.E)	Difference with Control (% inhibition)	Amount (S.E)	Difference with Control (% inhibition)	Amount (S.E)	Difference with Control (% inhibition)
00	1.03 (0.017)	-	0.13 (0.005)	-	0.59 (0.011)	-	1.57 (0.006)	-	0.43 (0.028)	-
100	1.02 (0.003)	0.01 (0.97)	0.14 (0.011)	0.01 (-7.6)	0.66 (0.017)	0.07 (-11.86)	1.06 (0.011)	0.51 (32.48)	0.42 (0.028)	0.01 (2.32)
250	1.02 (0.001)	0.01 (0.97)	0.11 (0.005)	0.02 (15.38)	0.68 (0.005)	0.09 (-15.25)	0.81 (0.028)	0.76 (48.40)	0.33 (0.008)	0.1 (23.25)
500	1.05 (0.005)	0.02 (-1.94)	0.12 (0.005)	0.01 (7.69)	0.65 (0.011)	0.06 (-10.16)	0.71 (0.023)	0.86 (54.77)	0.22 (0.028)	0.21 (48.83)
1000	1.04 (0.005)	0.01 (-0.97)	0.15 (0.017)	0.02 (-15.38)	0.63 (0.017)	0.04 (-6.77)	0.62 (0.013)	0.95 (60.50)	0.14 (0.006)	0.29 (67.44)
2000	1.01 (0.005)	0.02 (1.94)	0.18 (0.011)	0.05 (-38.46)	0.61 (0.017)	0.02 (-3.38)	0.31 (0.008)	1.26 (80.25)	-	-

Five amino acids i.e., L-leucine, L-methionene, DL (+) Alanine, glutamic acid and Aspartic acid were prominently detected in the control and treated maize seeds. Amino acids were found to be decreased with the increases in the concentrations of the toxin (AFT-B_1). The maximum inhibition in L-leucine and glutamic acid contents was 79.24 and 77.78% respectively at 2 ppm concentration of AFT-B_1, L-methionine, DL (+) Alanine and Aspartic acid were completely inhibited at that concentration of AFT-$B_{1.}$ Minimum inhibition 22.64, 20.00, 30.64, 43.13 and 14.28% was recorded in L-leucine and L-Methionine. DL (+) Alanine, glutamic acid and Aspartic acid due to 100 µg/l concentration of AFT-B_1, respectively (Table-01). Change in Amino acid due to treatments of various concentration of *Andrographis paniculata, Azadirachta indica* and *Adhatoda vasica* were also evident in Table -02, 03 and 04.

In our experimentation, a highly significant decrease in the levels of amino acids of maize seedlings has been observed with the treatment of different levels of aflatoxin B_1 and plant extracts separately (Tables- 01, 02, 03 and 04). The maximum inhibitions in L-leucine and glutamic acid have been found up to 79.24% and 77.78% at 2000 µg/l (2 ppm) concentration of AFT B_1, respectively. However, L- methionine, DL (-) alanine and aspartic acid were completely lost at that concentration (Table-10), the partial utilization of these amino acids in the germinating process. May be assumed and further strengthened on the basis of respiratory quotient values of the treated seeds, which have been found less than one in all the cases. R.Q. values below one indicated that the respiratory substrate was protein and not the carbohydrate or fat. Inhibition in proteins and amino acids levels suggested that these components are finally converted into keto acids which participated in the respiratory cycle.

Annapurana and Rao (1983) have, however, reported that the levels of amino acids continued to decrease as the concentration of the test chemical (Metacid 50) increased. Prasad *et. al.*, (2018 a & b) revealed that plant extracts also influenced the concentration of Amino acid contents in aflatoxin B_1 treated maize seeds.

CONCLUSION

This paper highlights recent research in Bihar that was designed to progress the monograph on aflatoxin contamination of maize using Bitter plants as a biological monitoring method with the target of reducing aflatoxin contamination. The results indicate that the biocontrol product has a high potential for reducing aflatoxin contamination in local environmental conditions.

REFERENCES

1. Cotty P.J., Bayman P., Egel D.S., Elias K.S. Agriculture, aflatoxins and *Aspergillus*. In: Powel K.A., Renwick A., Peverdy J.F. (1994). *The Genus Aspergillus: from Taxonomy and Genetics to Industrial Application.* Plenum Press; New York, NY, USA: Pp. 1-27.

2. Bandyopadhyay R., Ortega-Beltran A., Akande A., Mutegic., Atehnkeng J., Kaptoge L., Senghor A.L. Adhikari B.N., Cotty P.J. (2016). Biological control of aflatoxins in Africa: Current status and potential challenges in the face of climate changes. *World Mycotoxin J* : 1-20.

3. International Agency for Research on Cancer (IARC). (2002). *Monograph on the evaluation of Carcinogenic Risks to Humans. Some Traditional Herbal medicines, some mycotoxins, Naphthalene and styrene.* Volume 82. IARC; Lyon, France: Aflatoxins; pp. 171-300.

4. Madan, N. Medina, A. (2016). Integrating gene expression ecology and mycotoxin production by *Fusarium and Aspergillus* species in relation to interacting environmental factor. *World Mycotoxin Journal,* 9 (5), pp. 673-684.

5. Okafor, N. (2007). Palm Wine, In: Modern Industrial Microbiology and Biotechnology, Okafor, N. (ed). Science Publishers, USA, pp: 270-271.

6. Munaya, C. (2013). Bitter leaf-based extracts cures hepatitis co-inferation and other. *The Guardian Newspaper,* July, 25.

7. Sharma, P. and Sanadhya, D. (2017). The King of Bitters, "*Andrographis paniculate*". A Plant with Multiple Medicinal Properties. *The Journal of Plant Science Research 33 (1)*, 117.

8. Singh, A. Khan, M. M. A. Sahu, D. Vishwakarma, N. Yadav, A. and Singh, A. N. (2017). Pharmacological and Anti-bacterial Activities of the leaves of *Andrographis paniculate* Nees. *Journal of Pharmacognosy and Phytochemistry.* Vol. 6 (3), Pp. 418-420.

9. Sinha, K. K. Niranjan Kumar and Prasad, G. (1993). Use of mustard (*Brassica juncea L.*) and (*Cicer arietinum L.*) seedling germination inhibition assay for aflatoxin $B_{1.}$ *Mycopathologia, 121*(3): 175-178

10. Sidhu, O. P. Chandra, H. and Bhel, H. M. (2009). Occurrence of aflatoxins in mahua (*Madhuca indica Gmel.*) seeds: Synergistic effect of plant extracts on inhibition of *Aspergillus flavus* growth and aflatoxin production. *Food and Chemical Toxicology, 47:* 774-777.

11. Sinha, B. K. (2018). Studies on the Inhibitory Effect of some Plant Extracts on Mycoflora and Aflatoxin production. *ACTA Scientific Microbiology, 1.6:* 23-25.

12. Thanaboripat, D. Cheunoy, W. Peteharat, U. Ruangrattametee, V. and Kraisintu, K. (2000). Control of aflatoxigenic fungi by Thai neem. *Government Pharmaceutical Organization Journal,* 21, 41-49.

13. Thanaboripat, D. (2011). Control of Aflatoxins in Agricultural Products using Plant Extracts. *KMITL Sci. Tech. J. Vol. 11.*

14. Thanaboripat, D. Naranong, N. and Pccrapakorn, N. (1989). Effect of some herbs on growth of *Aspergillus flavus* and Aflatoxin Production. *Srinakarinwirot Journal of Science,* 5, 33-39.

15. Thanaboripat, D. (2003). Mycotoxins: occurrences and control in foods. In International Union of Food Science and Technology, ed. *The International Review of Food Science and Technology,* U. K: *IUFoST,* pp. 130-133.

16. Thanaboripat, D. Nontabenjawan, K. Leesin, K. Teerapiannont, D. Sukeharoen, O. and Ruangrauanametee, V. (1997). Inhibitory effect of garlic, clove and carrot on growth of *Aspergillus flavus* and aflatoxin production. *Journal of Forestry Research,* 8, 39-42.

17. Thanaboripat, D. Mongkontanawut, N. Suvathi, Y. and Ruangrattametee, V. (2004). Inhibition of aflatoxin production and growth of *Aspergillus flavus* by citronella oil. *KMITL Science Journal,* 4 (1), 1-8.

18. Ramachandran, J. and Sankaranarayanan, S. (2013). Antifungal Activity and the mode of Action of Alkaloid Extract from the Leaves of *Adhatoda vasica. International Journal of Ethnomedicine and Pharmacological Reasearch 1 (1),* 80-87.

19. Razzaghi-Abyaneh, M. Sharms- Ghahfarokhi, M. Yoshinari, T. Rezaee, M. B. Jaimand, K. Nagasawa, H. and Sakuda, S. (2008). Inhibitory effects of *Satureja hortensis* L. essential oil on growth and aflatoxin production by *Aspergillus parasiticus. International Journal of Food Microbiology,* 123, 228-233.

20. Reddy, K. R. N. Raghavender, B. N. Reddy, C. R. Salleh, B. (2010). Biological control of *Aspergillus flavus* growth and subsequent aflatoxin B_1 production in sorghum grain. *African J. of Biotech. 9*: 4247-4250.

21. Reddy, K. R. N. Nurdijati, S. B. and Sallch, B. (2010). An overview of plant-derived products on control of mycotoxigenic fungi and mycotoxins. *Asian Journal of Plant Science,* 9, 126-133.

6

A Comparative Review of Marine and Freshwater Diatoms Regarding Community Structure

Priti Kumari and K.K. Sahu

University Department of Botany,
L.N. Mithila University, Darbhanga, Bihar-846004

INTRODUCTION

Diatoms are abundant diverse and important components of the algal population in aquatic ecosystems. They can respond rapidly to environmental variations and are being increasingly used to assess water quality. The relation between diatoms and environmental variables is strong and quantifiable making diatoms appropriate quantitative indicators of ecological conditions in aquatic systems. Evaluation of the diatoms community concerning water quality may involve direct analysis based on main species or a more complex statistical approach using multivariate techniques.

The relation between diatoms and environmental variables are strong and quantifiable, making diatoms appropriate quantities indicators of ecological conditions in marine and freshwater. Remains of diatoms in marine (oceanic) and freshwater sediments are valuable indicators of past environments and have recently been used to detect PH changes. Deposit of sedimentary diatoms also called diatoms is important in many industries due to the fine structure and inert nature of the siliceous material.

Diatoms are one of the most beautiful microscopic creatures in the world. Diatoms are most beautiful like jewellery, despite the huge difference in the area of marine and freshwater ecosystems. They occur in both marine and fresh habitats where among other groups of algae, they play pivotal rates in food web and energy transport. Since diatoms are an integral part of the community where they live, it is often difficult to reparate their contribution to ecosystem functioning from the benefits of the mixed communities. However, the role of diatoms as "indicators" has been emphasized because they respond sensitively to changes in environmental conditions and reversal species have wide geographical distribution and relatively well-known ecological and relatively well-known ecological demands.

Diatoms have a significant role in global primary production, too marine environments are highly overrepresented in this aspect (Yool &Tyrrell,2003, Ragueneau et al 2006, Benoiston *et at* 2017.

MATERIALS AND METHODS

Bihar in general and North Bihar in particular, is very rich in inland water bodies samastipur is an important city of North Bihar. It is situated in at 25 latitude and longitude covering around 2900 sq KM. It has an average elevation of 56 meters. The city has many perennial ponds and an important river of North Bihar, Burhi Gundak passes through the city.

Five different experimental sites, three rivers Burhi Gandak and two ponds were selected for the study. The three sites of the rivers Burhi Gandak were Near Rahamatpur as site-I, under the bridge at a railway station as site-II and near the village Bahadurpur as site-III, two selected ponds were: Pond of B.R.B.College as pond-I and Mahadev pond at laguna as pond-II. Samples were collected from these five sampling sites on a seasonal basis from June 2015 to December 2016

RESULTS AND DISCUSSION

In the present study of diatom community structure, their density, species richness and species evenness were determined.

Table 1

Season	Year	River				
		Site I	Site II	Site III	Pond-I	Pond-II
Summer	2015	1100	1600	1200	2200	2100
	2016	1500	1200	1100	2700	2700
	2015	300	400	300	1500	1600
Rain	2016	500	300	400	1400	1400
Winter	2015	1600	400	2900	2000	2100
	2016	1800	2100	2300	1500	1600

Table 2

	Yearly mean		Seasonal means		
	2015	**2016**	**Summer**	**Rain**	**Winter**
River	1088.89	1244.44	1283.33	366.67	1850
Pond	1916.67	1883.33	2425	1475	1800

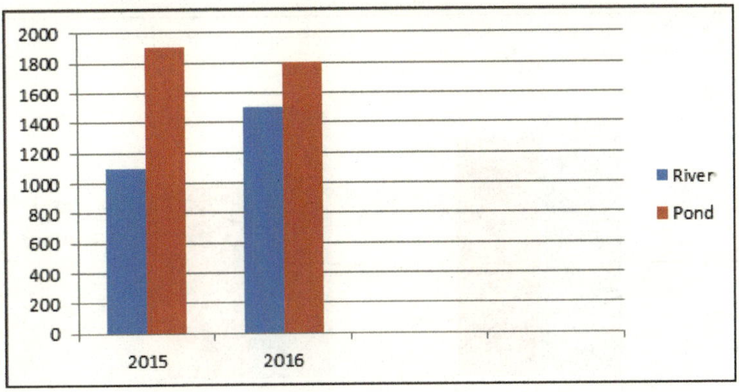

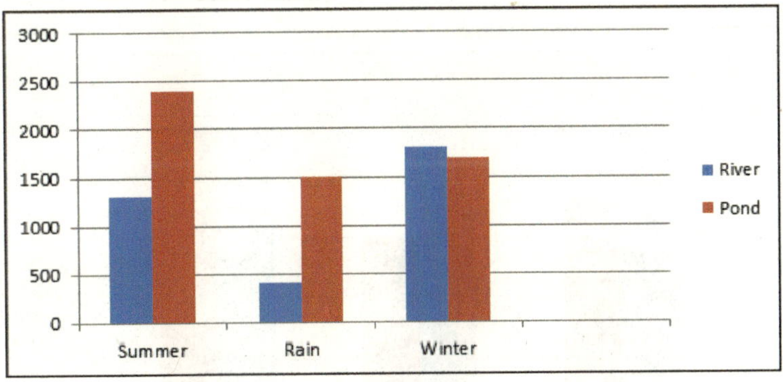

Table 3

Season	Year	River			Pond-I	Pond-II
		Site I	**Site II**	**Site III**		
Summer	2015	2.12	2.74	2.15	2.05	2.6
	2016	2.34	2.37	2.34	1.67	1.8
Rain	2015	1.97	2.05	1.87	1.57	1.72
	2016	1.85	1.91	2.12	1.32	1.4
Winter	2015	2.21	2.67	2.59	2.63	2.2
	2016	2.43	2.35	2.42	2.51	2.31

Table 4

	Yearly mean		Seasonal means		
	2015	2016	Summer	Rain	Winter
River	2.26	2.24	2.34	1.96	2.45
Pond	2.13	1.81	1.9	1.5	2.41

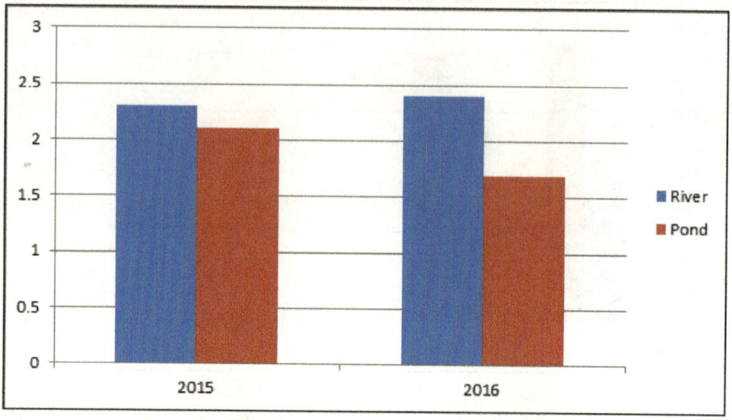

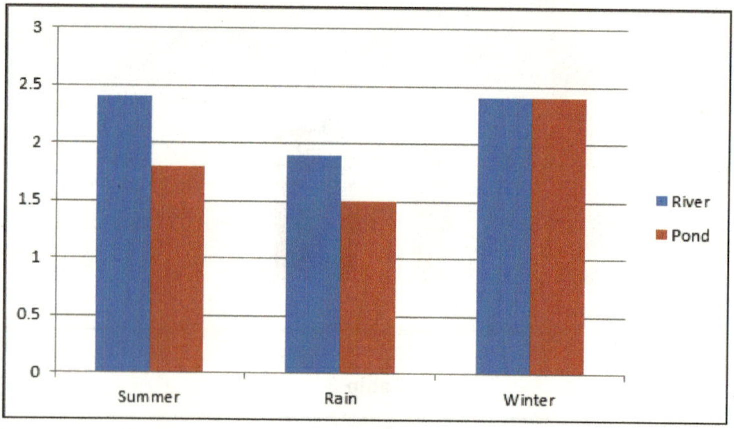

Table 5

Season	Year	River			Pond-I	Pond-II
		Site I	Site II	Site III		
Summer	2015	1.85	3.11	1.98	1.75	1.77
	2016	2.05	2.57	2.26	1.16	1.1
Rain	2015	1.73	2.34	1.43	1.16	1.21
	2016	1.65	1.97	1.75	0.81	0.85
Winter	2015	2.03	3.07	2.81	2.87	2.85
	2016	2.24	2.92	2.86	2.65	2.61

Table 6

	Yearly mean		Seasonal means		
	2015	**2016**	**Summer**	**Rain**	**Winter**
River	2.37	2.25	2.3	1.81	2.82
Pond	1.94	1.55	1.36	1.01	2.75

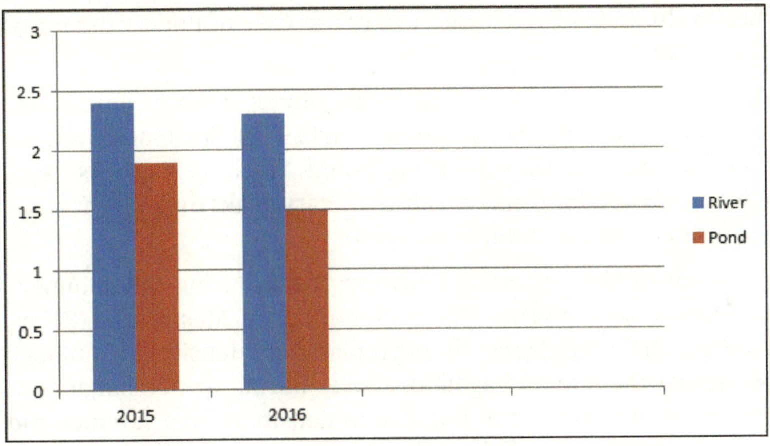

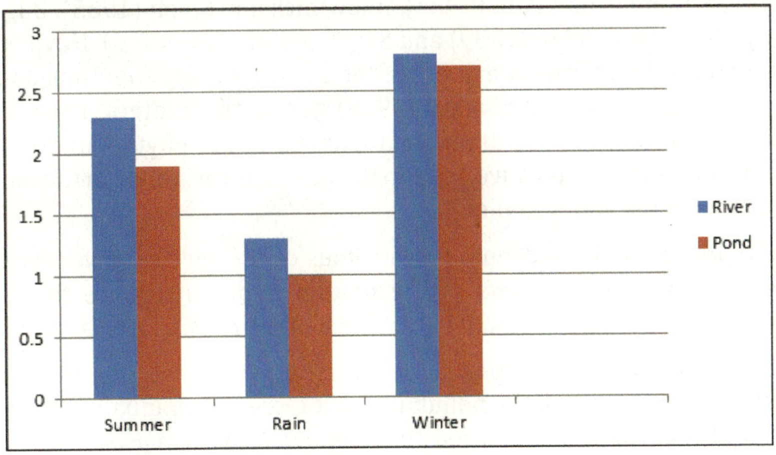

Diatoms represent outstanding bio indicators for the assessment of water quality. In the present chapter, Diatom community structure has been studied in terms of Density, species diversity, Species Richness and species Evenness. In the present study seasonal mean of the Density was maximum in winter and minimum in rain in river whereas the seasonal mean of the water density was maximum in summer and minimum in rain in case of pond. The maximum density of the diatoms during the winter has been reported by Dutta et al (1954) Ramkrishnaish and Sarkar(1982), Bhatt et al (1985) and Bilgrami et al (1985)

but Pahwa and Mehrotra (1076), Khan and siddiqu1(1974) and singh and Ahmad (1990) have reported summer dominance of the diatoms.

The phytoplankton algae play an important role in aquatic ecosystems. The phytoplankton population is mostly structured by the physical and chemical variables of their environment. The annual cycle of phytoplankton population build-up has been recorded almost on a bimodal pattern with a primary peak of higher magnitude during winter and a secondary peak of relatively lower height in summer in the case of river whereas in the case of pond primary peak was during the winter.

A bimodal pattern of fluctuation in the phytoplankton population density has also been reported earlier by Grorer et al (1978), Singh and Swarup (1979). Saha(1995) Verma and Munshi (1987) and Singh (1992). As regards the corresponding reasons for primary and secondary peaks the present observations confirm some of them and differ from others.

Graver et al (1978) reported a primary peak during the summer months and a secondary peak during the winter months. Mesfin and Belay (1989) have reported the abundance of phytoplankton density in summer which decreases during the rain and again increases during winter. Singh and Swarup (1979) recorded two maxima of population density during summer and winter respectively. Summer peak of phytoplankton population has also been reported by Rai(1978), Nandan and Patel(1984), Bhowmick and Singh (1985), Bilgrami *et al* (1985), Mesfin and Belay (1989) and Singh and Ahmad (1990). However, Vyas and Kumar (1968) reported peak of phytoplankton during the rains and Bhatt et al (1985) and Sengar and Sharma (1985) during the winters. The variations regarding the season of the minima and maxima of the phytoplankton density discussed above may be conceived due to the inherent variations in the biotic and abiotic factors of the water bodies.

In the present study seasonal mean values of the species. Diversity species Richness and species Evenness of the diatom population were found to be maximum in winter and minimum in rain both in the river as well as a pond.

There is an apparent reduction in the diversity values near the effluent out-fall irrespective of the nature of the pollutant but it increases gradually downstream. Woodwell (1970) has also reported that the diversity increased with distance from toxic pollution. The findings of most of the workers reveal that the diversity index declines with increasing pollution. But contrary results have also been reported, that is, an increase in diversity index in polluted water (Elaranta 1976, Trivedy 1980 and Trivedy et al 1985). This increase of diversity in polluted waters has been well explained by Trivedy (1980), According to him the behaviour of the diversity index with eutrophication type of pollution is different. It is quite probable that eutrophication may lead to quite probable that eutrophication may lead to increase in the number of species as well as the number of individuals increasing and reaching the level of pollution, then the diversity index decreases remarkably because of an increase in the number of individuals of only a few pollution tolerant species. Clean water, which are usually oligotrophic, will have lower diversity values compared to eutrophic waters, but with a higher level of pollution the diversity index declines. However, there will always be a decline in the diversity index in a toxic type of pollution.

It is evident from the available literature that the diversity index changes differently in different tupe of pollution. Therefore, this index can be used on the one hand to determine the extent of pollution and on the other to determine the type of pollution as well. Some workers have proposed scales of diversity indices applied to the population of benthic macroinvertebrates in a stream receiving organic waters. Wilhm and Dorris (1968) proposed a scale based on their work on benthic macroinvertebrates, and staub et al(1970)proposed a scale based on the diversity of the phytoplankton.

Wilhm and Dorris (1968) have classified the water bodies witnessing diversity less than I as heavily polluted, between 1 and 3 as moderately polluted, and exceeding 3 as clean water.

Species diversity of the river water ranged from 1.85 to 2.74 and 1.32 to 2.51 in pond water. Therefore, it may be inferred on the basis of classification proposed by wilhim and Dorris (1968) that the river and pond water is moderately polluted. Further, the values of species diversity suggest that the river and pond water is less polluted in summer and winter compared to rain. Although water bodies with higher diversity generally indicate good quality of water a low diversity does not necessarily indicate poor quality of water. It can be substantiated by the findings of Hawker (1979), that the low diversity index of an aquatic body is not always due to the physical conditions prevailing at that experimental site

CONCLUSION

The Diatoms community structure and content of silicate of five experimental water bodies three of the river Burhi Gantak (lotic ecosystem) and two o ponds (lentic ecosystem) at samastipur, Bihar were studied during two consecutive years, Density of diatom, species Diversity, Species Richness and Species Evenness were found comparatively. Pond water showed higher. The content of silicate in pond water was observed higher than that of river water except in the rainy season. A significant positive correlation between silicate with species diversity and species richness was found. A mass culture of diatoms and its role in the prophylactic and stress management of local crops is suggested. The bioassay technique can be used to strengthen this suggestion.

REFERENCES

1. APHA (2017). Standard Methods for the Examination of Water and waste Water 23rd Edn. American Public Health Association Washington, 1-1504.

2. Biligrami K.s Munshi Dutta J.S. Siddiqui E.N and Singh N.K. (1979). Primary productivity of phytoplankton of the river Ganga, *Biol Bull India*,1,39-42

3. Bhatt S.D Bisht Y and Nagi U. (1985), Hydrology and phytoplankton population in river kosi of Western Himalaya (Uttar Pradesh) *Indian J.Ecol*,12(1) 141-146.

4. Chaudhary A.N (2015). Chemical Science Physico- Chemical Analysis of Soil Collection from Mahishi, Dist Saharsa (Bihar). *International Journal of Chemical Sciences*,13(2),1034-1038.

5. Clarke G.L (1965) Elements of Ecology. Pul. John Wiley and Sons, Inc., New York London, Sydney, 560.

6. Claude E.B (2014). Silicon, Diatoms in aquaculture. Global Aquaculture Advocate,38-39.

7. Chow T (1958). A study of water quality in the fish ponds of Hong Kong, *Univ. Fish. J.* 2,7-28

8. Das S.M and Pandey J (1976) Some physico-chemical and biological indicators of pollution in lake Nainital, Kumaum (UP), *Indian Journal of Ecology,* 5(1) 7-16

9. Ducan S.W., and Blinn D.W. (1989).Importance of physical variables on the seasonal Dynamics of Epilithic Alage in a highly Shaded Canyon Stream. *J. Phycol.*, 25(3),455-461.

10. Govindan V.S and Sundaresan B.B (1979). Seasonal succession of algal flora in a polluted region of Adyar river. *Indian J, Environ Health*, 21(2),131-142

11. Hosmani S.P (1988). Seasonal changes in phytoplankton communities in a freshwater pond at Dharwar Karnataka state, *India Phykos*, 27(land2), 82-87

12. Hosmani S.P and Bharti S.G (1980) Limnological studies in ponds and lakes of Dharwad-Comparative phytoplankton ecology of water bodies. *Phykos*, 19(1),27-43

13. Khan M.M. and Senaya G. (1985). Ecology of planktonic blue-green algae in the profiles of industrially polluted Husainsagar Lake Hyderabad, India phykos, 24(1), 140-148

14. Kant S. and Anand V.K (1978). Interrelationship of phytoplankton in sewage fed pond, Phykos: *Journal of the Phycological Society Ecology,* 28(2), 137-146

15. Kumar A. and Bohara C (2002). Faunistic composition community structure, population dynamics and ethology of the birds of Udhuwa lake (central pargana) and the strategies for their conservation of lakes, reservoir and rivers (Ed.A.Kumar) ABD Publishers, Jaipur,149-199.

16. Kant s. and Anand V.K (1978). Interrelationship of phytoplankton and physical factors in Mansard Lake, Jammu (J and k) *Ind. J. Ecol*, 5(2), 134-140.

17. Muntazuddin S, Azad A.K, Bharti P. and Ranjan R (2012) Physico–chemical analysis of groundwater of the Bhdhi Ganddak belt in Muzaffarpur district, India. *International Research Journal of Environment Sciences*, 1<1>,7-11
18. Maps of India (2012). https://w.w.w.mapsofindia.com/maps/bihar/rivers/ (Accessed (02/09/2017)
19. Mesfin M. and Belay A(1989). A study of seasonal variation of phytoplankton in relation to water quality in Legedadi Reservoir. *Int. J.Ecol. Environ.Sci,* 15, 1-16
20. Ministry of Works and Housing Drinking Water Standards (1975). Chemical and biological methods for water pollution studies, by Trivedi, R.K and Goel, P.K (1984), Environmental Publications Karad, Maharastra, India,2020.
21. Moyle J.B (1949). Some indices of lake productivity. *Trans. Amer. Fish Soc,* 76,322-334
22. Mahadev J. and Ahmed A.S (2009). Physico-chemical factors influencing the growth of diatoms in two habiotats of Mysore. *Asian Journal of Environmental Science,* 4(1), 12-14.
23. Mallik S. and Bose S.K (1987), Ecological study of planktonic. Blue Geen Algae of fresh water pond of Ranchi, Bihar. *Indian Int. J Ecol. Sci.* 13,149-152.
24. Nasar S.K.J and Nasar S.A.K. (1978). Primary productivity of phytoplankton in sewage-fed pond. Phykos: *Journals of the Phycological Society* (India), 17(land 2),51-54.
25. Prasad B.N and Singh Y. (1982) on diatom as an indicator of water pollution. *J. Indian Bot Soc,* 61,326-336.
26. Philipose (1959). Freshwater phytoplankton of inland fisheries Proc. Symp Algology, 4 241-262.
27. Ray D. Rao R.R Bhoi A.V, Biswas A.K Ganguly A.K and sanyal P.B (2000). Physico-chemical Quality of Drinking Water in Rohtas District of Bihar. Environmental monitoring and Washington, 1-1504, ISBN: 978-0-87-553-287-5.
28. Reddy M.P and Venkateswarlu V.(1987) .Water quality and pollution in the river Tunga bhadranear Kurnool. Andhra Pradesh. *J. Environ Biol,* 8(2),109-119
29. Rana B.C and Palria S. (1988). Phycological and physicochemical evaluation of the river Ayad, Udaipur *Phykos* (Algiers),27(1), 211-217
30. Singh S.R Swarup K. (1979) Limnological studies of Suraha Lake (Baillia) II. Periodicity of phytoplankton *J. Indian.Bot. Soc.* 58(4),319-329.
31. Seulthope C.D (1967) Biology of aquatic vascular plants. Edward Arnold (Pub) Ltd. London 10.

32. Surabhi(1994), Limnological studies of Dighi pond at Darbhanga (Bihar). Phycological Laboratory, University Department of Botany, B.R.A Bihar University, Muzaffarpur,1-198.

33. Sabater S, Sabater F and Tomas X (1987), Water quality and diatom communities in two Catalan rivers (N.E.Spain) *Wat. Res,* 21(8), 901-911,

34. Sengar R.M.S , Sharma K.D and Pathak P.D (1985), Studies on the distribution of algal in polluted and non-polluted regions in Yamuna river at Agra (U.P). *J.Ind. Bot.Soc,* 64(4),365-376.

35. Singh A.K (1985) Physico-chemical and bacterial study of sewage water discharged into the river Ganga at Bhagalpur, India. Environmental and Ecology, 3(2),138-142.

36. Sahia R, Saxena P.K and Jabeen S.(1985), Ecological survey of the algal flora of polluted habitats of Gorkhpur ,*Phykos* (Algiers), 24(1),4-11.

37. Singh V.P. Tewari A and Khan M.A (1970). Studies on seasonal variation of algae of sewage. *Phykos,* 9, 57-62.

38. Somashekar R.K. (1984). Contribution of the algal flora of river Kapila, Karnataka, II Diatoms Hykos, 23(1-2),125-129.

39. Venu P. Kumar V, Sardana R.K and Bhasin M.K (1984). Indicatory and functional role of phytoplankton in effluents of rangpo distilleries of Sikkim Himalayas, *Phykos,*23(1,2) 38-44.

40. Verma P.K and Munshi J.D (1987). Plankton community structure of Badua reservoir, Bhagalpur, Bihar. *Trop. Ecol.* 28,200-207.

7

Exploration of Various Medicinal Parasitic Plants on the Diverse Host in the North Bihar Region of India

Rajnigandha and Khwaja Salahuddin*

University Department of Botany,
L. N. Mithila University, Darbhanga - 846 004, Bihar, India
**Corresponding author: ksalahuddin@lnmu.ac.in, salahuddin212@gmail.com*

INTRODUCTION

Parasitic plants are traditionally very important for medicinal properties in the Northern region of Bihar. They are mostly in the family of Convolvulaceae, Loranthaceae, Orobanchaceae and Viscaceae. *Cuscuta* (dodder), is also known as aakashbel or amarbel in vernacular terms. (Chen *et al*; 2013). According to older literature, it is an angiosperm parasite belonging to the family Convolvulaceae the most common species are *Cuscuta reflexa*, *Cuscuta chinensis,* and *Cuscuta campestris*. Its geographical distribution, wide host range and difficulty in control make *Cuscuta*, one of the most harmful parasites present worldwide (Parker and Riches, 1993; Kaiser, 2015). *Cuscuta* is the most noticeable among all parasites. It has so many hosts belonging to dicots, but monocots are less preferred (Nawakocha and Aigbokan, 2013; Rowntree *et al.*, 2014). The members of the *Monogynella* subgenus are fruit killers that can kill and damage fruits, but the species of subgenus *Cuscuta* prefer herbaceous hosts and are more harmful than *Monogynella* as in the case of the subgenus, *Grammica.* Parasitic plants have many activities, and the most reported and pronounced effects are antibacterial, anti-inflammatory, antioxidant, anticancer and neuroprotective properties (Palambo, 2009; Chepkuri, 2020). The scientific literature regarding the traditional uses has revealed remarkable.

These parasites are a bizarre group of plants with countless chemical properties yet are poorly understood in many areas (Ghimire *et al.*, 2004; Joshi *et al*; 2011)). The name of the plant for which the medicine is prescribed, the family, the use of the part, the disease, the method of administration, etc. are mentioned. Although these plants are very common in several parts of the Mithila region and widely studied from the ethnomedicinal point of view but the current study primarily focuses on the ethnomedical perspective of local rural communities Riches and Parker (1995) pointed out that Broomrape species are related to crops in the Apiaceae, Asteraceae, Brassicaceae, Cucurbitaceae, Fabaceae and Solanaceae families and are therefore directly involved in food consumption by humans. Like many weeds, some pathogenic plant parasites are spread. Ethnobotanical information is an indispensable tool for drug development. Overall, the ethnobotanical use of parasitic plants is important. Medicinal parasitic plants in rural areas of developing countries are the major source of medicines for human use (Palambo, 2009). However, the traditional knowledge of these plants is seriously threatened by the current trends of intensive agriculture, industrialization, and economic globalization that is promoting the migration of rural people to urban areas (Bhogankar and Deokule, 2002; Bhattarai *et al*; 2009).

MATERIALS AND METHODS

The study was conducted in the Mithila region of Bihar. It consists of Darbhanga, Madhubani and Samastipur districts of Bihar [Map 1]. The meteorological data and physical features of the study areas were collected year-wise and district-wise, respectively.

Several species of *Cuscuta, Cassytha, Loranthus, Viscum, Striga, Orobanche and Castillega* were collected in the study areas. The species diversity was investigated, and all major clades were identified (Mc Neal *et al*, 2013). For taxonomic studies, parasitic plants were collected from several habitats of all three districts. The plant materials were collected in flowering seasons and complete notes were made. The collected plants were identified, noted family-wise and described in detail.

Several field visits were conducted during the research period and ethnobotanical data were collected through interviews, and structured questionnaires from various traditional healers, herbalists, and other elders of the Mithila region to find out the efficacy of parasitic plants and their parts used in the treatment of various diseases and disorders. In this regard, regular visits were made to different sites of Darbhanga, Madhubani and Samastipur districts to collect samples of parasitic plants and further to observe all hosts infected with different parasitic plants.

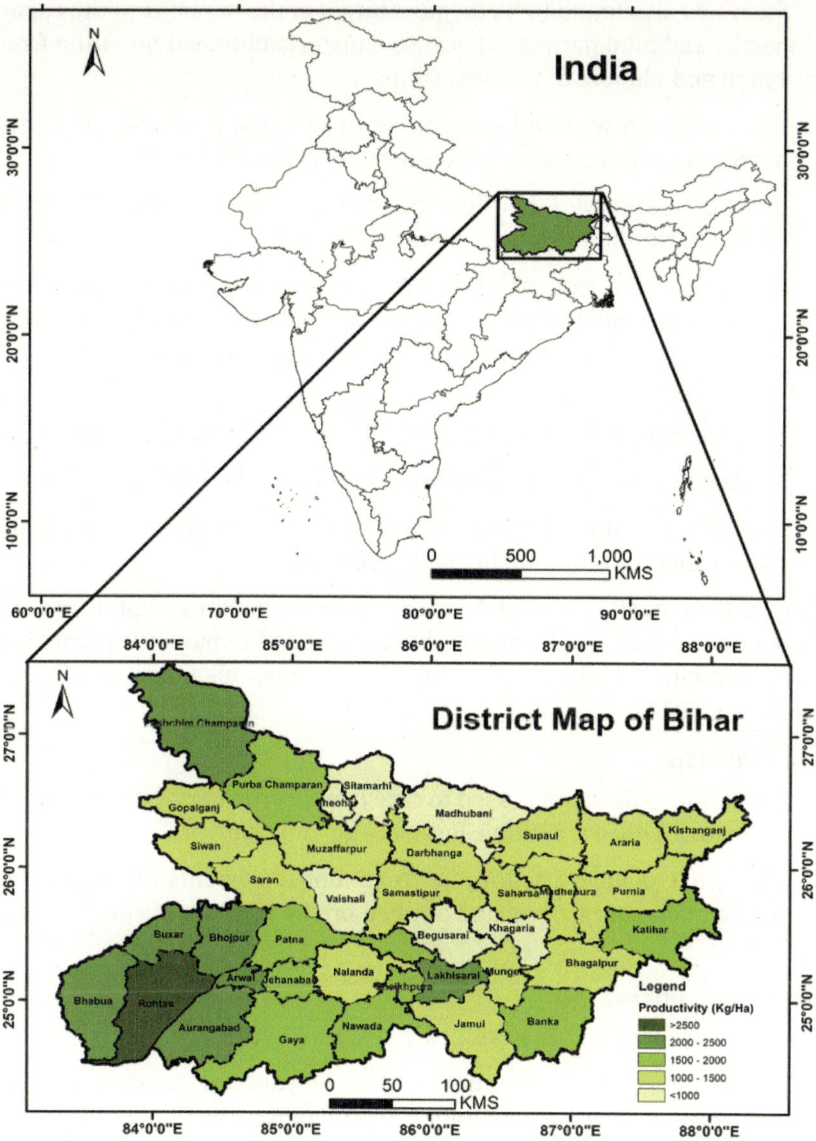

Map1: Location map of Bihar

RESULTS AND DISCUSSION

Findings:

- It was noticed that all parasitic plants were found growing on whole plants of the host because of the development of haustoria which obtained nutrition from all parts.

- Stem was also found to be the next target of the parasitic plant, especially partial and total parasites because haustoria obtained nutrition from the xylem and phloem of the host stem.
- The leaf was also found susceptible to parasitic plants because they get nutrition from the leaf much easier than other plant parts.
- It was also observed that flowers, fruits, and seeds provided nutrition to the parasitic plants to some extent.
- It was reported from the literature and personal investigation that the leaf was the most important plant part showing medicinal properties (Rokaya *et al.*, 2010; Singh *et al.*, 2012; O'Neil and Rana, 2017).
- It was observed that decoction of whole plant, bark, flowers and fruits of parasitic plants showed healing properties in several ailments.
- Fruits and seeds of parasitic plants also showed medicinal properties.
- Seed was found to be not as essential as other parts of parasitic plants, with respect to ethnomedicinal properties.

Many of the parasites are used as folk medicine in many regions around the Mithila region of Bihar. The bioactive chemicals found in parasitic plants include phenolic compounds, especially flavonoids, glycosides, alkaloids, and fatty acids (Tahir and Gul, 2014)

Cuscuta chinensis:

- Decoction of the stem is used to cure jaundice, urinary disorder problems, diarrhoea, cholera, asthma, fever, cough and cold.
- The whole plant is used in the treatment of ailments like baldness. The whole plant was ground and applied on the head for curing hair fall esp. in men

Orobanche aegyptiaca;

- It is used to treat kidney stones and scabby ulcers
- Decoction is used as a herbal tonic meant for treating impotence, spermatorrhea and dermatological problems.

Viscum album:

- Decoction of young twigs and leaves are used in the treatment of respiratory ailments. The paste of tendril is applied for insect bites.
- The paste of the plant is administered to patients to reduce fever, and in the treatment of ulcers, blood disease, fractured bone, dislocation, and cancerous wounds (Abdel-Salam *et al.*, 2010).

Loranthus odoratus:

- Leaves are used in the treatment of hypertension, diabetes, and rheumatism and possess immunomodulatory, hypolipidemic properties, antimicrobial and antioxidant properties (Yusuf et al. 2013).

- Fruits are used in indigestion.

Dendropthoe falcata:

- Decoction of the whole plant is used to treat bronchitis and respiratory troubles.
- Decoction of bark is the remedy for asthma and menstrual disorders in women. The bark is used in impotency.

Flowers of *Cuscuta chinensis*

Orobanche aegyptiaca

Dendrophthoe falcata

Loranthus odoratus

Viscum articulatum

CONCLUSION

The findings of the study confirm and validate the great potential and ethnomedical use of parasitic plants. A great proportion of parasitic plants have been used as folk medicine in various regions around the Mithila region of Bihar. Based on the ethnobotanical survey, the leaf was found to be showing maximum medicinal properties. The species of *Cuscuta* were found to be medicinally important while others are threats to natural ecosystems and crops. These plants are harmful to the growth of the host but are beneficial as well. These plants must be explored and harnessed up to a significant level in order to find out the medicinal importance of their specific parts.

REFERENCES

1. Abdel-Salam, O.M., Sleem, A.A., Shaffie, N.M. (2010). Effect of *Viscum album* on acute hepatic damage caused by carbon tetrachloride in rats. *Turkish Journal of Medical Science*. 40:421-426.

2. Bhattarai, S., Chaudhary, R.P., and Taylor, R.S.L. (2009). Ethnomedicinal plants used by the people of Nawalparasi district, Central Nepal. Our Nature.

3. Bhogaonkar, P.Y.and Deokule,S. (2002). Some useful ethnomedicinal plants of Korku of Melghat region [MS], *Ethnobotany,* 12.

4. Chang, M., and Lynn, D.G. (1986). The haustorium and the chemistry of the host recognition in parasitic angiosperm. *Journal of Chemical Ecology*, 12: 561-579.

5. Chen, G. T., Lu, Y., Yang, M., Li, J. L., and Fan, B. Y. (2018). Medicinal uses, pharmacology, and phytochemistry of Convolvulaceae plants with central nervous system efficacies: A systematic review. *Phytotherapy Research, 32*(5): 823-864.

6. Chepkirui, W. (2020). Effects Of Climate Variability on Dodder Invasion, Distribution and Management in Belgut Area of Kericho County, Kenya (Masters dissertation, Kenyatta University).

7. Ghimire, S.K., Mckey, D., and Aumeeruddy-Thomas, Y. (2004). Heterogeneity in ethnoecological knowledge and management of medicinal plants in the Himalayas of Nepal: implication for conservation. *Ecological Society.*

8. Ibrar, M. (2002). Responsibilities of ethnobotanists in the field of medicinal plants. In Proceeding of Workshop on curriculum development in the applied ethnobotany. Ethnobotany Project, WWF Pakistan, 34-D /2, pp:16-20

9. Joshi, K., Joshi, R., Joshi, A.R. (2011). Indigenous knowledge and uses of medicinal plants in Macchegaun, Nepal. *India Journal of Traditional Knowledge,*

10. Joshi, G.C.P., and B.P. Kothyari. (1985). The new host of *Dendrophthoe falcata* (Linn.f.) Etting. *Indian Journal of Forester,* 8:235.

11. Kaiser B., Vogg, G., Furst, U.B., Albert, M. (2015). Parasitic plant of genus *Cuscuta* and their susceptible and resistant host plant. *Frontiers in Plant Science,* 6:45. doi:10.3389/fpls.2015.00045

12. Mahwasane, S.T., Middleton, L., and Boaduo, N. (2013). An ethnobotanical survey of indigenous knowledge on medicinal plants used by the traditional healers of the Lwamondo area, Limpopo province, South Africa. *South African Journal of Botany,* 88: 69-75.

13. Nwakocha, M.I and Aigbokhan, E.I. (2013). Host range and preference of *Cuscuta campestris* [Yunck) among common weeds in Benin city, Nigeria. *Nigerian Journal of Botany,* 26 (2): 1-29.

14. O'Neil, A., and Rana, S. (2017). An ethnobotanical analysis of parasitic plants in the Nepal Himalaya. *Journal of Ethnobiology and Ethnomedicine,*12(14):14.

15. Palombo, E. A. (2009). Traditional medicinal plant extracts and natural products with activity against oral bacteria: potential application in the prevention and treatment of oral diseases. *Evid-Based Complementary and Alternative Medicine,* 2011, 1-15.

16. Parker, C. and Riches, C.R. (1993). *Cuscuta* species, the dodders; and *Cassytha filiformis.* In. Parasitic Weeds of the Worlds: Biology and Control (eds C. Parker and CR Riches), 183 223. CAB International, Wallingford, UK.

17. Rokaya, M.B., Munzbergova, Z., Timsina, B. (2010). Ethnobotanical study of medicinal plants from the Humla district of Western Nepal. *Journal of Ethnopharmacology.*

18. Rowntree, J.K., Barham, D.F., Stewart, A.J.A., and S.E. Hartley. (2014). The effect of host species on a keystone parasitic plant and its aphid herbivores. *Functional Ecology,* 28(4): 829–836.

19. Singh, A.O., Kumar, A., and Terai, D.D. (2012). An ethnobotanical survey of medicinal plants used in Terai forest of western Nepal. *Journal of Ethnobiology and Ethnomedicine.*

20. Singh, N.P., Khanna, K.K. and Mudgal, V. (2001). Flora of Bihar: Analysis (Ist Ed.). Botanical survey of India.

21. Tahir, I., and Gul R.K, M. (2014). The study of the elemental profile of some important medicinal plants collected from distract Karak, Khyber Pakhtunkhwa, Pakistan. *International Journal of Chemical and Pharmaceutical Science,* 6: (5).

22. Verma, N., & Yadav, R. K. (2018). *Cuscuta reflexa*: a paracitic medicinal plant. *Plant Arch, 18,* 1938-1942.

23. Yusuf, L., Oladunmoye, M.K., Ogundare, A.O., Akinyosoye, F.A., Daudu, O.A.Y. and Hassan, G.A. (2013). Antimicrobial and antioxidant properties of mistletoe (*Viscum album)* growing on cola (*Cola nitida*) tree in Akure North, Nigeria. *Journal of Microbiology Research and Reviews,* 1(3): 35-41.

8

Biotechnology in Food and Health Security: A Review

Maryam Parween and Arindam Kumar

Dept. of Botany, J.M.D.P.L. Mahila College, Madhubani, Bihar

INTRODUCTION

Biotechnology, particularly in the realms of food and health security, has emerged as a transformative force, offering innovative solutions to some of the most pressing challenges facing humanity today. As the global population continues to grow, the demand for sustainable food production and improved health outcomes has never been more critical. Biotechnology provides powerful tools, such as genetic engineering, to enhance food production, nutrition, and healthcare.In the context of food security, biotechnology enables the development of genetically modified crops that are more resistant to pests, diseases, and environmental stresses. These advancements not only increase agricultural productivity but also contribute to the resilience of food systems, reducing the risk of food shortages and malnutrition. Furthermore, biotechnological innovations such as biofortification have made it possible to enhance the nutritional content of staple crops, addressing micronutrient deficiencies in vulnerable populations.In the realm of health security, biotechnology plays a pivotal role in the development of new therapies, vaccines, and diagnostic tools. Genetic engineering, in particular, has revolutionized medicine by enabling the creation of targeted treatments for diseases, personalized medicine, and advanced gene therapies. These innovations have the potential to significantly improve healthcare outcomes, reduce disease burdens, and increase life expectancy.

However, the integration of biotechnology in food and health security is not without challenges. Ethical considerations, regulatory frameworks, and public acceptance remain key issues that need to be addressed to ensure the safe and equitable application of these technologies. As we move forward, a balanced and informed approach is necessary to harness the full potential of biotechnology for the benefit of society, while mitigating potential risks.

MATERIALS AND METHODS

The research methodology for biotechnology in food and health security involves a combination of experimental, analytical, and interdisciplinary approaches to develop, assess, and implement biotechnological innovations. The methodologies can be broadly categorized into the following phases:

The methodology for addressing biotechnology's role in food and health security involves a multidisciplinary approach integrating scientific research, technological innovation, policy analysis, and implementation strategies. Below is a detailed narrative of the methodological framework. Biotechnology is revolutionary for agriculture and the food system, because it places control over food production in the genes. Food production, of course, has always been empowered by genes, but we haven't been able to see them, precisely select them, or move them across traditional species barriers. Now we can. And day by day we are learning which traits in crops and livestock are controlled by individual genes, how to turn those genes on and off, how to splice them into the organisms, and how to amplify gene products.

So first, we have an awesome new technology that operates at the genetic level of the food system--the most fundamental level of food characterization. This means that the production and quality commands in the food system begin with the genes and, most importantly, with those who hold the genes and wield the new genetic technologies.

The first phase involves identifying and developing biotechnological solutions that address specific challenges in food and health security. This includes:**Genomic Analysis**: Sequencing and analyzing genomes of crops, livestock, and pathogens to identify traits that improve productivity, disease resistance, or nutritional value. **Synthetic Biology**: Engineering novel genetic circuits and synthetic pathways to enhance biological functions in organisms, such as creating probiotics for health improvement or biofortified crops.**Fermentation Technology**. Developing microbial systems for producing alternative proteins, vitamins, and essential nutrients. **Vaccine Development:** Using recombinant DNA technology to create effective and scalable vaccines against emerging diseases. By systematically following these methodological steps, biotechnology can be harnessed effectively to enhance food and health security, ensuring a sustainable and equitable future.

RESULTS AND DISCUSSION

Biotechnological tools such as CRISPR and polymerase chain reaction (PCR) have revolutionized disease diagnostics by providing rapid and accurate results. During the COVID-19 pandemic, these technologies played a crucial role in testing and tracking the virus, enabling timely public health responses. Biotechnological advancements have also supported the development of biofertilizers, biopesticides, and nitrogen-efficient crops, promoting sustainable agriculture. For example, nitrogen-fixing crops reduce the need for synthetic fertilizers, which are often associated with greenhouse gas emissions and water pollution. By integrating these technologies, farmers can achieve higher productivity while minimizing environmental impacts.

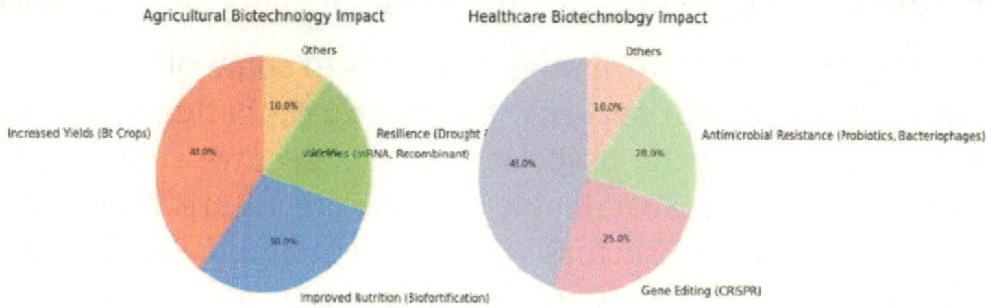

Fig.1 Pie charts illustrating the impact of biotechnology in food and health security

Biotechnology often addresses food and health security simultaneously. For example, functional foods, developed through genetic engineering, provide both nutritional benefits and disease prevention. Probiotic-enriched dairy products and omega-3-enhanced crops are examples of foods that promote health while combating malnutrition.

Another example is the use of edible vaccines, where genetically modified crops are engineered to produce antigens. These vaccines can be consumed directly, providing immunization in regions with limited access to healthcare infrastructure.

The integration of biotechnology into food and health systems has far-reaching socioeconomic and environmental implications. Biotechnology plays a pivotal role in enhancing crop yields through genetic modification. Genetic modification allows plants to be engineered for higher productivity and resilience. For instance, the insertion of specific genes can enable crops to utilize nutrients more efficiently, thereby promoting vigorous growth and higher yields. This genetic intervention can also shorten growth cycles, allowing for multiple harvests within a single growing season. One significant advantage of biotech innovations is the development of disease-resistant crops. By introducing genes that confer resistance to specific pathogens, biotechnologists have created varieties that can withstand attacks from viruses, bacteria, and fungi. This not only reduces the dependency on chemical pesticides but also ensures stable food production in areas prone to plant diseases. The Papaya Ring spot Virus-resistant papaya is a notable example, which saved the papaya industry in Hawaii. Beyond yield increases, genetic modification also improves the nutritional profiles of crops. Biofortified crops, such as Golden Rice, have been engineered to contain higher levels of essential vitamins and minerals, such as vitamin A, iron, and zinc. These nutritionally enhanced crops can help combat malnutrition in regions where nutrient deficiencies are prevalent, thereby contributing to overall food security.

Moreover, biotech advancements support sustainable farming practices that are essential in the face of climate change. Techniques such as no-till farming, facilitated by herbicide-resistant crops, maintain soil structure, reduce erosion, and enhance carbon sequestration. These methods not only improve soil health but also contribute to long-term agricultural sustainability and resilience.

Biotechnology offers transformative solutions to some of the most pressing challenges in food and health security. By fostering a better understanding of its benefits and addressing the associated challenges, we can harness its full potential to create a more sustainable and secure future. The collaborative efforts of stakeholders across sectors will be essential in achieving these goals. While biotechnology offers immense potential, it is not without challenges. Public concerns about the safety and ethical implications of GMOs and biopharmaceuticals must be addressed. Misinformation and resistance to biotechnology can hinder adoption, particularly in regions where public trust in science is low. Biotechnology contributes to environmental conservation by reducing reliance on harmful agricultural inputs and promoting efficient resource use. Herbicide-resistant crops enable no-till farming, which preserves soil health and reduces carbon emissions. Additionally, biotechnological solutions help mitigate the impacts of climate change on agriculture and health systems.

CONCLUSION

In conclusion, biotechnology, through the tools of genetic engineering and other advancements, stands at the forefront of enhancing food and health security. Its contributions are vital in ensuring a sustainable and healthy future, making it an indispensable component of modern science and technology. Biotechnology stands at the forefront of efforts to achieve global food and health security. Through its ability to enhance agricultural productivity, combat malnutrition, prevent and treat diseases, and promote environmental sustainability, biotechnology offers transformative solutions to some of humanity's most pressing challenges. By addressing ethical concerns, ensuring equitable access, and fostering innovation, biotechnology can pave the way for a future where food and health security are universally attainable. Farmers adopting GMOs often experience increased income due to higher yields and reduced costs associated with pesticides and fertilizers. These benefits are especially impactful in developing countries, where agriculture forms the backbone of the economy. Biopharmaceuticals and advanced diagnostics have also driven down healthcare costs by enabling early disease detection and more effective treatments, reducing hospital stays and long-term care expenses.

REFERENCES

1. B Yousuf, K Gul, A A Wani ,P Singh (2015). Health benefits of anthocyanins and their encapsulation for potential use in food systems: A review. Crit. Rev. Food Sci. Nutr. 56: 2223-2230.

2. C Umetsu (2022). Sustainable farming techniques and farm size for rice smallholders in the Vietnamese Mekong Delta: A slack-based technical efficiency approach. Agric. Ecosyst. Environ. 326: 107775.
3. C Vogel , E.M Marcotte (2012). Insights into the regulation of protein abundance from proteomic and transcriptomic analyses. Environmental genetic engineering: Hope or hazard Curr. Sci. 70(9): 823-832.
4. H W. Plaxton. Lambers (2014) Sensing signaling and control of phosphate starvation in plants—molecular players and application in Phosphorus Metabolism in Plants. Rev. 48: 25-64.
5. J Robinson (2013). Breeding the nutrition out of our food. Online. New York Times. Royte, Elizabeth. The Post-GMO Economy. Modern Farmer, December 6
6. J Wang , J Sun , J Miao , J Guo , Z Shi , Y Chen , X Zhao , B Li , F Han , Y Tong , A Li Z (2013). Phosphate starvation response regulator TaPHR1 is involved in phosphate signaling and increases grain yield in wheat. Ann. Bot. 111: 1139-1153.
7. P Salvi , R Agarrwal , N Kajal Gandass , M Manna, H Kaur , R Deshmukh (2022) "Synthetic biology approaches to engineering the nitrogen symbiosis in cereals" J. Exp. Bot. 65: 1939- 1946.
8. S M Whitney , RE Sharwood , D Orr , SJ White , H Alonso , J Galmés (2011). Isoleucine 309 acts as a C4 catalytic switch that increases ribulose1, 5-bisphosphate carboxylase/oxygenase (rubisco) carboxylation rate in Flaveria. Proceed. Nat. Acad. Sci. 108(35): 14688- 14693.
9. Statista (2018). Monsanto's net sales in the seeds and genomics segment from 2011 to 2017 by specialty (in million U.S. dollars).
10. TA Shaikh, WA Mir, T Rasool , S Sofi. (2022). Machine learning for smart agriculture and precision farming: Towards making the fields talk. Arch. Comput. Methods Eng. 1-41.
11. WR Scheible, M Rojas-Triana (2015). Sugar transporters and their molecular tradeoffs during abiotic stress responses in plants. Physiol. Plant. 174(2): e13652.
12. X Wang , Y Wang , J Tian , BL Lim , X Yan , H Liao (2009). Over expressing At PAP15 enhances phosphorus efficiency in soybean. Plant Physiol. 151: 233-240.

9

Biocontrol Potential of Indigenous Lab against *F. oxysporum* Causing Wilt of Tomato in Samastipur

Shikha Raj and Shambhu Nath Jha

Dept. of Botany, K.S.R. College, Sarairanjan, Samastipur, Bihar

INTRODUCTION

Tomato (*Solanum lycopersicum* L.) is one of the most important horticultural crops and economically significant and nutritionally vital crops worldwide – wide belongs to the Solanaceae family, contributing to global food security and dietary diversity. Tomatoes are cultivated basically in two main climates: temperate and tropical areas, which have allowed the tomato to be one of the most grown vegetables worldwide. Although tomatoes are widely grown in several countries, the place of production includes; China, India, United States, Turkey, Brazil, Iran. Tomatoes are a rich source of essential nutrients such as vitamins A and C, potassium, and lycopene—a powerful antioxidant linked to various health benefits. Their culinary versatility ranges from raw consumption in salads to being transformed into sauces, soups, juices, and sundry processed products. *F. oxysporum* f. sp. *lycopersici*, responsible for vascular wilt disease, several measures are used to control tomato diseases caused by *Fusarium* species, where the chemical fungicides are the most used, but they need to be replaced due to their harmful effect on the environment and human health (N. Zebboudj, *et al.*2020). Different alternatives strategies can be used, and biological control is the most important alternative. Important groups of microorganisms used in the biological control of fungal diseases are lactic acid bacteria (LAB), that are commonly found in fermented food products such as milk, yogurt, cheese, wine, meat and soil. LAB are capable of synthesising various forms of organic acids, bacteriocins and hydrogen peroxide which is known to hinder pathogenic fungi and bacteria.LAB attribute has been identified to have some positive characteristics inclusive of synthesis of antibiotics, improvement of plant growth and nutrient uptake, and strengthening the plant's immune system(Steglińska *et al.,* 2022). The Indigenous LAB isolated from the rhizosphere of healthy plants are preferred since they are suited for the local geophysical environment. *Fusarium* wilt is a soil-borne disease which is caused by this pathogen attacks the tomato plants in the vascular tissue resulting to wilting of the plants and death. The disease is prevalent in Samastipur region of North Bihar and has resulted in decline in yield. *Fusarium* species are also the causal organism of *Fusarium* foot rot and *Fusarium* seedling blight of small grain cereals resulting in reduced germination and poor plant establishment leading to intensified yield and quality losses(Byrne *et al.,* 2022).This research aims to investigate the biocontrol potential of indigenous LAB strains isolated from the Samastipur region against *Fusarium oxysporum* and their efficacy in managing tomato wilt under field conditions.

MATERIALS AND METHODS

Isolation and Identification of Pathogens

Tomato samples were interviewed from areas of Samastipur in North Bihar and test soils were taken from each individual plot.Isolation of pathogens from the infected part of Tomatofruits and leaves from several locations in Samastipur and brought these in sterilized polybag to the laboratory immediatelyforisolation (Figure 1). I have used PDA (potato dextrose agar) media on sterilized Petri plate for inoculation of diseased area at temperature 26^{0}C±(optimum temp.) for seven days. Then cultured from infected tomato plants exhibiting a normal wilt syndrome.*F. oxysporum* culture suspension was prepared by washing 10 days old Culture of *F. oxysporum* in a Petri dish with spore germination solution.

LAB strains were grown on MRS (de Man, Ragosa and Sharpe) agar and grown anaerobically at 37°C for 48 hr. Spread the enriched cultures onto MRS agar plate and then choose the typical LAB colonies. The isolated LAB colonies were subculture and stocked for another investigation. The biochemical tests, and the 16S rRNA sequencing were used to identify the isolated LAB strains. Microscopical and biochemical characterization involving Gramstaining, catalase activity and characterization of carbohydrate fermentations was done on the LAB isolates.

Figure 1. Tomato plant showing wilt symptoms in leaves and fruit.

In Vitro Antagonistic Assay

The bio-chemical antagonism of the LAB isolates against *F. oxysporum* was evaluated through the dual culture method. *F. oxysporum* grow as a mycelial disc was placed at middle of the PDA plate while LAB strains were cultured around the disc at four different radial positions. After that plates were incubated at 25 °C for seven days. Control of fungal growth was determined by comparing the diameter of the colony of *F. oxysporum* in the presence and absence of LAB.Percentage of growth inhibition (GI) was calculated using the formula: GI (%) = C0 – CF / Co x 100%, where C0 is the dry weight of fungal mycelium (control), CF is the dry weight of fungal mycelium after inhibition by LAB.

Statistical Analysis

The data were analyzed using ANOVA andsignificant differences between treatments were determined using Duncan's multiple range test at a significance level of $p < 0.05$.

RESULTS AND DISCUSSION

Out of all the soil samples, 15 LAB strains were isolated. Therefore, from the data obtained from the morphological, biochemical, and molecular tests, the isolates belonged to the genera *Lactobacillus, Pediococcus,* and *Leuconostoc*(Table 1).The results of the dual culture assay showed that several LAB strains had high antagonistic tendencies against the fungus *F. oxysporum*.Among all tested strains *Lactobacillus plantarum* and *Pediococcuspentosaceus* demonstrated the highest inhibition zones and decreased fungal growth of 65% and 70% respectively in comparison with the control.The antifungal activity of *P. pentosaceus* culture supernatant was determined to be at the stationary phase near the end of exponential growth phase and was also found to be dependent on the pH.The regulators of the metabolites against the fungi also exhibited stability to heat and resistance to enzymatic digestion by proteases. Substances that possess antifungal characteristics were found in the culture fractions ranging from 500 to 1400 Da.

Table 1 Biochemical characters of isolated lactic acid bacteria

Strain	Gram Stain	Catalase Test	Oxidase Test	Gas from Glucose	Growth at 15°C	Growth at 45°C	Lactic Acid Production
LAB01	+	–	–	–	+	–	High
LAB02	+	–	–	+	+	+	Moderate
LAB03	+	–	–	–	+	–	High
LAB04	+	–	–	+	–	+	Moderate

Strain	Gram Stain	Catalase Test	Oxidase Test	Gas from Glucose	Growth at 15°C	Growth at 45°C	Lactic Acid Production
LAB05	+	–	–	–	+	–	High
LAB06	+	–	–	+	+	+	Moderate
LAB07	+	–	–	–	–	+	High
LAB08	+	–	–	+	+	–	Moderate
LAB09	+	–	–	–	+	–	High
LAB10	+	–	–	+	+	+	Moderate
LAB11	+	–	–	–	+	–	High
LAB12	+	–	–	+	+	+	Moderate
LAB13	+	–	–	–	–	+	High
LAB14	+	–	–	+	+	–	Moderate
LAB15	+	–	–	–	+	–	High

The results on fig.2 show that mycelium growth decreased significantly in presence of LAB cells, compare to control plates on PDA culture media.The LAB treatment has resulted in approximately **29.4% growth inhibition** of *Fusarium spp.* after 7 days of incubation on PDA medium. The control plate shows robust fungal growth, while the test plate indicates restricted growth due to LAB's inhibitory effects.

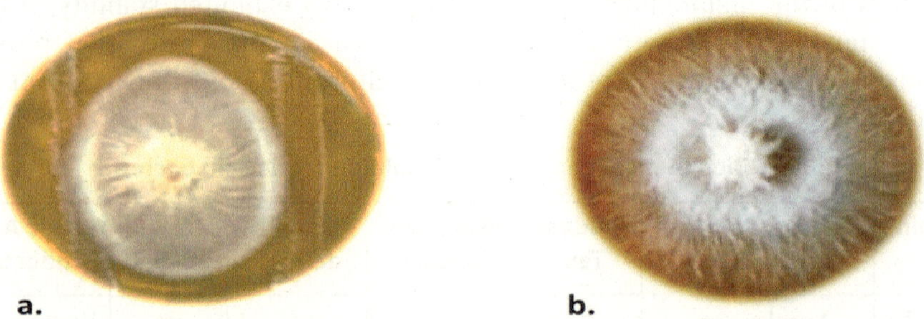

Figure 2. a. Growth inhibition of WY05 by Lld after 7-day incubation on PDA medium; b. Control plate.

The overlay assay analysis of the present study revealed variable response to the impact of LAB on Fusarium growth (Table 2).*L. plantarum* (Lmd) shows the strongest inhibitory potential, with complete inhibition of fungal growth (> 8% of plate area) across almost all isolates. Lmd exhibited substantial (P <

0.05) antifungal activity ranged from 4.3-19.7% while Lld had moderate (P < 0.05) inhibition from 2.1-7.9%, though Ldb only had non-significant (P > 0.05) inhibition 2.1% after 48 hours but reduced to 1% on the third day of incubation The present results clearly indicate that compared to the other strains, Strain WY18 was more sensitive to the action of all LAB used in this study. Contrary to its sensitivity to the other tested LABs, Strain WY03 was sensitive only to Lmd and resistant to Lp in this study.

Table. 2 Overlay assay of LAB on *Fusarium* spp. growth: +++=no growth in >8% of PCA; ++=no growth in between 3 to 8% of PCA; + = no growth in 0. 1 to 3% of PCA; – = no visible inhibition.

LAB Isolate	WY01	WY02	WY03	WY05	WY06	WY10	WY12	WY13	WY18	Control
L. plantarum (Lmd)	+++a	++b	+++	+++	+++	+++	+++	+++	+++	–
L. rhamnosus (Lld)	+c	++	–d	++	++	++	++	++	+++	–
L. brevis (Ldb)	+	–	–	+	+	+	+	+	++	–

Table 3 Identification of species and pathogenicity percentage of *Fusarium* isolates

Specimen Voucher	Species	Patho-genicity (%)	Signi-ficance	GenBank Accession No.	Similarity with GenBank Nucleotide Database (%)
F. oxysporum					
WY01	*F. oxysporum*	100%	a	MG973088	99.85%
WY03	*F. oxysporum*	100%	a	MG973090	99.85%
WY18	*F. oxysporum*	100%	a	MG973096	100%
F. solani					
WY02	*F. solani*	50.00%	b	MG973089	100%
WY10	*F. solani*	36.66%	b	MG973093	100%
WY12	*F. solani*	43.33%	b	MG973094	99.85%
F. redolens					
WY05	*F. redolens*	76.66%	c	MG973091	99.85%
WY06	*F. redolens*	80.00%	c	MG973092	99.85%
WY13	*F. redolens*	46.66%	c	MG973095	99.85%

The antifungal activity of LAB-CFS to inhibit *Fusarium sp.* CID124-CS mycelial growth using PDA method is showed in Table 3. The results obtained in this study showed that inhibitory activity was significantly different (P≤0. 05) against all *Fusarium sp.* CID124-CS after one week incubation period. Thus, it has been noted that the LAB-CFS of *P. pentoceous*1-MSS isolated from the soil strongly has been reduced the spreading of the *Fusarium sp.* CID124-CS about 60.0% after 7 d incubation at room temparature 28^0 C. However, others LAB-CFS were showed good reduction and lowest reduction was noticed around 35.0 mm with CFS of *Lb. Plantarum*1-MSS against *Fusarium sp.* CID124-CS as described in Table 4.

Table 4 Percent inhibition of mycelial growth of *Fusarium oxysporum* CID124-CS after 7 days of incubation at room temperature (28°C) using LAB-CFS on PDA.

LAB Isolate	Source	Inhibition (%)	Significance (a-e)
Lb. Plantarum1-MSS	Indigenous (Samastipur)	35.0	e
P. pentosaceus1-MSS	Indigenous (Samastipur)	60.0	a
Lb. acidophilus ATCC314	Reference Strain	40.0	b
Lb. plantarum ATCC8014	Reference Strain	38.0	c
Lb. Plantarum1-FCF	Indigenous (Samastipur)	36.0	d

The application of microbes aimed to suppress pathogens has been considered as a natural and environmentally friendly method of disease control.Due to these properties LAB are one of the most attractive candidates for a biological control because they are able to produce a lot of antifungal agents such as organic acids, hydrogen peroxide and bacteriocins.Considering this experiment, laboratory isolated bacteria (LAB) from soil has successfully exhibited disease control in tomato by minimizing the effects of fungal pathogens.

CONCLUSION

Tomato production may be limited by several phytopathogenic fungi and bacteria which caused yield and fruit quality losses. As a result of preparing high quality food, the need to protect tomatoes from pathogen is possible through the adoption of other methods of controlling the disease.This work demonstrates the biocontrol efficiencies of the indigenous LAB strains obtained from Samastipur against *F. oxysporum* that affects tomato through causing wilt disease.The LAB isolates from the soils samples was demonstrated to have similar antifungal activity as Lb. acidophilus ATTCC314, Lb. plantarum ATCC8014 and fermented foods isolateLb. Plantarum1-FCF.All the above LAB strains exhibited marked antifungal activity against tomato fungal wilt and could be potentially used for organic biological control. Organic acids, hydrogen peroxide and bacteriocins synthesised by LAB acted very effectively in preventing the pathogen fungus.This approach offers a sustainable alternative to chemical fungicides, reducing their harmful environmental impact.

REFERENCES

1. Antifungal activity of lactic acid bacteria against Fusarium species responsible for tomato crown and root rots. (2020). *Environmental and Experimental Biology*, 18(1). https://doi.org/10.22364/eeb.18.02

2. Akaram Husain, Zaiton Hassan, Saleh, A., W El-Mabrok, Mohd Nizam, and Mohd Bodrul Munir. (2019). In Vitro Efficacy of Lactic Acid Bacteria with Antifungal Activity Against Fusarium Sp. CID124- CS Isolate from Chilli Seeds. *Unpublished*. https://doi.org/10.13140/RG.2.2.29566.33606

3. Byrne, M. B., Thapa, G., Doohan, Fi. M., and Burke, J. I. (2022). Lactic Acid Bacteria as Potential Biocontrol Agents for Fusarium Head Blight Disease of Spring Barley. *Frontiers in Microbiology*, 13, 912632. https://doi.org/10.3389/fmicb.2022.912632

4. Gwiazdowski, R., Kubiak, K., Juś, K., Marchwińska, K., & Gwiazdowska, D. (2023). The Biocontrol of Plant Pathogenic Fungi by Selected Lactic Acid Bacteria: From Laboratory to Field Study. *Agriculture*, 14(1), 61. https://doi.org/10.3390/agriculture14010061

5. Jaffar, N. S., Jawan, R., and Chong, K. P. (2023). The potential of lactic acid bacteria in mediating the control of plant diseases and plant growth stimulation in crop production—A mini review. *Frontiers in Plant Science*, 13, 1047945. https://doi.org/10.3389/fpls.2022.1047945

6. Krishnan, S. V., Nampoothiri, K. M., Suresh, A., Linh, N. T., Balakumaran, P. A., Pócsi, I., and Pusztahelyi, T. (2024). Fusarium biocontrol: Antagonism and mycotoxin elimination by lactic acid bacteria. *Frontiers in Microbiology*, 14, 1260166. https://doi.org/10.3389/fmicb.2023.1260166

7. López-Seijas, J., García-Fraga, B., Da Silva, A. F., and Sieiro, C. (2019). Wine Lactic Acid Bacteria with Antimicrobial Activity as Potential Biocontrol Agents against *Fusarium oxysporum* f. Sp. Lycopersici. *Agronomy*, 10(1), 31. https://doi.org/10.3390/agronomy10010031

8. Shafi, Z., Ilyas, T., Shahid, M., Vishwakarma, S. K., Malviya, D., Yadav, B., Sahu, P. K., Singh, U. B., Rai, J. P., Singh, H. B., and Singh, H. V. (2023). Microbial Management of Fusarium Wilt in Banana: A Comprehensive Overview. In U. B. Singh, R. Kumar, & H. B. Singh (Eds.), Detection, Diagnosis and Management of Soil-borne Phytopathogens (pp. 413–435). *Springer Nature Singapore*. https://doi.org/10.1007/978-981-19-8307-8_17

9. Steglińska, A., Kołtuniak, A., Motyl, I., Berłowska, J., Czyżowska, A., Cieciura-Włoch, W., Okrasa, M., Kręgiel, D., and Gutarowska, B. (2022). Lactic Acid Bacteria as Biocontrol Agents against Potato (Solanum tuberosum L.) Pathogens. *Applied Sciences*, 12(15), 7763. https://doi.org/10.3390/app12157763

10. Vargas, C., Jiménez, H. R., Almario, C. G., and Almario, A. G. (2023). Effect of lactic acid bacteria on the control of *Fusarium oxysporum* and Ralstonia solanacearum on singly infected and co-infected tomato plants. *Journal of Applied Microbiology,* 134(2), lxac053. https://doi.org/10.1093/jambio/lxac053

10

Study of Alien Herb Plants and Phytochemical Screening

Aleya Siddiquee, Motilal Srivastava and Md. Sarfaraz Ahmad

University Department of Botany, Jai Prakash University, Chapra, Bihar

INTRODUCTION

Alien species are defined as those species that is introduced intentionally or unintentionally to any region, location or area which is not its natural habitat. According to the Convention on Biological Diversity (CBD) "Alien species" refers to a 'species, subspecies or lower taxon, introduced outside its natural past or present distribution; including any part, gametes, seeds, eggs, or propagules of such species that might survive and subsequently reproduce'. Some alternate terms used for alien species are non-native species, non-indigenous species and introduced exotic, feral, foreign, ornamental or weedy species. The International Union for Conservation of Nature (IUCN) defined IAS is an alien species which becomes established in natural or semi-natural ecosystems or habitats, is an agent of change, and threatens native biological diversity. Invasive aliens plant species are defined as naturalized species producing a large number of offsprings and at considerable distances from the parent plants with the potential to spread over a large area. These are also called exotic species which are plants, animals, pathogen and other organisms that are not natural in certain ecosystem. IAS treat to reduce natural Vegetation. A.S Raghubanshi et al., (2005), defined alien species as non-native or exotic organisms that occur outside their natural adapted ranges and dispersal potential. Many alien species support our farming and forestry systems in a big way. According to Reddy et. al. 2008, an alien plant also referred to as exotic, introduced, foreign, non-indigenous or non-native, is one that has been introduced by humans intentionally or otherwise through human agency or accidentally from one region to another. An alien plant that has escaped from its original ecosystem and is reproducing on its own in the regional flora 5 considered a naturalized species. M. Sarfaraz et al., (2012) worked on conservation and exploration of medicinal plants in addition to alien plants of Gopalganj.

Moony and Hobbs (2000) have explained the invasive alien (exotic) plants as the global extent and rapid increase in invasive species is homogenizing the world's flora and fauna and recognized as a primary cause of global biodiversity loss. Plant invasions have been recognized as one of the most serious global processes impacting the structure, composition and function of natural and semi-natural ecosystems (Moony and Hobbs, 2000). Dutta and Mukherjee (1916) reported exclusive medicinal uses of alien invasive plant species of Dumka, Jharkhand. Chatterjee (1940) reported 6850 species as endemic to India for the first time. After critical analysis of flora of India, it has been convincingly concluded by phytogeographers that India has a flora of its own as many as 6850 (33%) taxa of Indian flora are endemic to present Indian boundaries.

There are four main stages of invasion: introduction, naturalization, colonization and dispersal. Invasive plant species differ from native species in different morphological characteristics such as specific leaf area, seed size, height, flowering duration where invasive species showed crucially dominant in the investigated characteristics. Some invasive species adapt easily to a variety of habitat, and have high reproductive potential and impact on ecosystem. They successfully over-compete with local plant species, change the composition and structure of plant communities, repress seed germination and seedlings development of local native species. In some agriculture, IAPs can lead to the loss of crops production. In past decades, the risks associated with invasive species have increased because anthropogenic activities altering the environment have escalated rapidly. The introduction of alien species is directly or indirectly with anthropogenic, the species being introduced deliberately or accidentally.

According to the Biodiversity strategy for 2030, managing IAPs is a priority for the European Union (European Commission 2020).

Invasive plants species can be spread in the following ways:

1. Contaminated seed, feed grain, hay, straw, mulch etc.
2. Movement of unclean equipment and vehicles across uncontaminated lands.
3. Livestock and wildlife.
4. Spreading gravel, and road fill that contains seed.
5. Nursery industry.
6. Recreation.
7. Water and wind transportation.

MATERIALS AND METHODS

Though studies deal with the impact of alien species are increasing, but still literature on association of broad quantitative synthesis and factors responsible i.e., attributes of recipient and dynamic ecosystems and of the invaders themselves (Levine et al., 2003) are under progress. We thoroughly studied the different areas like Hathwa, Thawe, Bhore and Kuchaikote of District Gopalganj and observed a large number of invasive herb plants and their extensive use by local communities for medical purposes.

The geographical area of Gopalganj is approx 2033 Sq.Km. with 14 blocks and 1534 villages. It majorly receives rainfall from south-west monsoon with temperature (°C), relative humidity (%) and total Rainfall (mm). During this survey, regular visits were organized to the selected areas of Gopalganj followed by the meeting to local residents and respondents. The information collected from the respondents were the local name of the plant, their usage, habit, habitat and families. Participants consent was taken into consideration and data was recorded only from participants who were willing to do so. After specific identification and study, the plants were dried and pressed under a herbarium press. Before pressing, the plants were treated with 2% mercuric chloride solution to disinfect the plant material. The specimens after drying were mounted on the herbarium sheets of standard size (41.6 x 29.5 cm).

Phytochemical analysis based on extractions. Extraction was done by Soxhlet with methanol as the solvent (Cowan, 1999). The experimental plant *Ageratum conyzoides* (L.) was collected in the stage of flowering or fruiting from their natural habitat in the months of March and April from the different places like Hathwa, Thawe, Bhore and Kuchaikote of district Gopalganj, Bihar, India. The samples were kept in plastic zipped pouch and carried to laboratory for analysis. The study used whole plant, washed under running water, then rinsed with distilled water and then dried in shade for 15 days at room temperature. The dried plant was then powdered in an electric blender. Identification and authentication of each plant species was done by a plant taxonomist using customary methods **(Lawrence, 1951)** and each plant specimens were dried and submitted to the Department of Botany, Jai Prakash University, Chapra.

RESULTS AND DISCUSSION

The proposed study is based on **survey-based analysis** and **experimental analysis.** The first section *i.e.,* **Survey based analysis** involves the collection of invasive alien plant species. The information collected from traditional practioners, forest dwellers, vaids, hakims, old man, farmers, domestic curers and housewives was recorded. These plant species are enlisted in the table which was studied in detail to record their botanical name, common name or local name, name of family, their habit/life form, parts of plants used, uses and nativity. The second section *i.e.,* is composed of an **experimental analysis** of one selected invasive alien species from this proposed area. The aim of second section is to find out major active components in the species and thus determine the biological activities possessed by extracted active components. The findings of the study have eventually found some economically benefits.

The participants possessed very deep knowledge of the plants and uses of their individual parts. We observed in the results that most of the plants had medicinal role and every part role in some or other way. To summarize this observation, we mentioned all the concerned information in table.

Table: List of IAS plants under consideration for the survey

S.N.	Name of Alien Species	Common or Local name	Family	Habit/ Life form	Parts Used	Use	Nativity
1	*Ageratum conyzoides* L.	Jangli pudina	Asteraceae	Annual/Herb	Whole plant	Medicinal and Pharmacological uses	Tropical America
2	*Alternanthera sessilis* (L.) R.Brazil. ex DC.	Garundi	Amaranthaceae	Annual or perennial/Herb	Stems and Leaves	Fodder/Agroforestry and Medicinal uses	Tropical America
3	*Amaranthus spinosus* L.	Kanta chaulai	Amaranthaceae	Annual/Herb	Seed, stems and leaves	Medicinal uses	Tropical America
4	*Anagallis arvensis* L.	Krishnaneel	Primulaceae	Annual/Herb	Leaves	Medicinal uses	Europe
5	*Argemone mexicana* L.	Phirangi dhatura	Papaveraceae	Annual-perennial /Herb	Seed and leaves	Medicinal uses	South America
6	*Blumea eriantha* DC.	Buradi	Asteraceae	Perennial/Herb	Whole plant, flowers and seeds	Medicinal uses	Tropical America
7	*Blumea lacera* (Burm.f.) DC.	Jangli Muli	Asteraceae	Annual/Herb	Whole part	Medicinal and edible uses	Tropical America
8	*Cannabis sativa* L.	Bhang	Cannabaceae	Annual/Herb	Bark, fruit and leaves	Medicinal uses	Central Asia
9	*Cassia hirsute* L.	Woolly cassia	Caesalpiniaceae	Annual-perennial /Herb	Leaves, seeds and roots	Edible, Medicinal and Agroforestry uses	Tropical America

S.N.	Name of Alien Species	Common or Local name	Family	Habit/ Life form	Parts Used	Use	Nativity
10	*Cassia occidentalis* L.	Kasunda	Caesalpiniaceae	Annual/Herb	Seeds, leaves and roots	Medicinal, Pharmaceutical and Agroforestry uses	South America
11	*Cassia pumila* Lam.	Sarmal	Caesalpiniaceae	Annual-perennial /Herb	Leaves, seeds and roots	Medicinal uses	Tropical America
12	*Cassia tora* L.	Chakunda	Caesalpiniaceae	Annual-perennial /Herb	Whole plant, roots, leaves and seeds	Edible and Medicinal uses	South America
13	*Catharanthus roseus* G.Don	Sada bahar	Apocynaceae	Perennial/Herb	Roots and leaves	Medicinal uses	Tropical America
14	*Celosia argentea* L.	Garke	Amaranthaceae	Annual-perennial /Herb	Leaves, roots and whole plants	Edible and Medicinal uses	Tropical America
15	*Chenopodium album* L.	Bathuwa	Chenopodiaceae	Annual/Herb	Whole plant and seeds	Edible and Medicinal uses	Europe
16	*Cleome viscosa* L.	Bagra	Capparidaceae	Annual/Herb	Leaves and seeds	Medicinal and Agroforestry uses	Tropical America
17	*Corchorus olitorius* L.	Pat	Tiliaceae	Annual/Herb	Leaves and seeds	Edible and Medicinal uses	Tropical Africa
18	*Cynodon dactylon* (Linn.) Pers	Doob grass	Poaceae	Perennial/Herb	Whole plant and leaves	Edible, Medicinal and Agroforestry uses	Africa

S.N.	Name of Alien Species	Common or Local name	Family	Habit/ Life form	Parts Used	Use	Nativity
19	*Cyperus cyperoides* L.	Kode jhar	Cyperaceae	Perennial/Herb	Leaves and Roots	Edible, Medicinal and Commercial uses	Tropical America
20	*Echinochloa colonum* (L.)Link	Jangli Jhangora	Poaceae	Annual/Herb	Seeds	Edible uses	Tropical South America
21	*Echinochloa crus-galli* (L.)Beauv.	Samak	Poaceae	Annual/Herb	Seeds, young shoots, stems and roots	Edible, Medicinal and Agroforestry uses	Tropical South America
22	*Echinops echinatus* Roxb.	Oont kanteli	Asteraceae	Annual/Herb	Leaves, roots and seeds	Medicinal uses	Afghanistan
23	*Eclipta prostrata* (L.)	Madhukar	Asteraceae	Perennial/Herb	Leaves, young shoot, stems, roots and flowers	Edible and Medicinal uses	Tropical America
24	*Europephorbia hirta* L.	Bara dudhi	Euphorbiaceae	Annual/Herb	Aerial parts	Medicinal uses	Tropical America
25	*Galinsoga parviflora* Cav.	Pardesi	Asteraceae	Annual/Herb	Leaves, stems and flowering shoots	Edible, Medicinal and Agroforestry uses	Tropical America

S.N.	Name of Alien Species	Common or Local name	Family	Habit/ Life form	Parts Used	Use	Nativity
26	*Gomphrena globosa* Linn.	Gul-e-makhmal	Amaranthaceae	Annual/Herb	Flowers and whole plant	Medicinal uses	Tropical America
27	*Hyptis suaveolens* (L.) Poit.	Vilaiti tulsi	Lamiaceae	Annual/Herb	Seeds, leaves and roots	Edible, Medicinal and Commercial uses	South America
28	*Ipomoea carnea* Jacq.	Behaya	Convolvulaceae	Annual/Shrub	Leaves and whole plant	Edible, Medicinal, Agroforestry and Ornamental	Tropical America
29	*Ipomoea eriocarpa* R.Brazil.	Bhawar	Convolvulaceae	Annual/Herb	Leaves, seed and roots	Edible, Medicinal and Agroforestry uses	Tropical Africa
30	*Ipomoea pes-tigridis* L.	Panchpatia	Convolvulaceae	Annual/Herb	Leaves, roots and seeds	Medicinal uses	Tropical Africa
31	*Ipomoea purpurea* (Lin.) Roth.	Glory	Convolvulaceae	Annual/Herb	Seeds	Medicinal uses	America
32	*Mimosa pudica* L.	Chui-mui	Mimosaceae	Annual-Perennial /Herb	Leaves, seeds and flowers	Edible, Medicinal and Agroforestry uses	Brazil
33	*Mirabilis jalapa* L.	Gul abbas	Nyctaginaceae	Perennial /Herb	Leaves, seeds and roots	Edible, Medicinal and Commercial uses	Peru

S.N.	Name of Alien Species	Common or Local name	Family	Habit/ Life form	Parts Used	Use	Nativity
34	*Nicotiana plumbaginifolia* Viv.	Ban tamaku	Solanaceae	Annual /Herb	Leaves	Medicinal uses	Tropical America
35	*Ocimum americanum* L.	Kali Tulasi	Lamiaceae	Perennial /Herb	Leaves and seeds	Edible and Medicinal uses	Tropical America
36	*Oxalis corniculata* L.	Amrul	Oxalidaceae	Annual-perennial /Herb	Whole plant, leaves and flowers	Edible and Medicinal uses	Europe
37	*Portulaca quadrifida* L.	Chicken weed	Portulacaceae	Annual/Herb	Seeds and leaves	Medicinal uses	Tropical America
38	*Solanum nigrum* L.	Mokoi	Solanaceae	Annual/Herb	Fruits, leaves and whole plant	Edible and Medicinal uses	Tropical America
39	*Tridax procumbens* L.	Kanphuli	Asteraceae	Perennial/Herb	Leaves	Edible, Medicinal and Agroforestry uses	Tropical America
40	*Xanthium strumarium* L.	Chota dhatura	Asteraceae	Annual/Herb	Leaves, roots and fruits	Edible and Medicinal uses	Tropical America

Natural active components from medicinal plants are important source of medicines for fatal diseases. They are better replacements of chemical agent which have several side effects on the user. Therefore, besides recording the invading capacity of the plant, there usages were also determined based on the participant's information.

Primary Screening for Phytochemical Compounds

Phytochemicals of *Ageratum conyzoides* was screened for alkaloids, flavonoids, glycosides, saponins, tannins, terpenoids, phenols and anthraquinone. The standard methods of these phytoconstituents were followed given by **Wall** *et al.,* (1952), Harborne (1984), Sofowora (1993), Trease and Evans (2002) and Junaid *et al.,* (2006).

Active Component Analysis using GC-MS: Gas Chromatography was used to separate the compounds present in the extract and identified based on Mass Spectrometry. Plant extracts were analyzed using GC-MS technique.

The plant extract of *Ageratum conyzoides* (L.) L. (ACE) showed identified compounds were propanephosphonic acid, bis- (trimethylsilyl) ester (RT:10.83 min), Fluoren-9-ol, 3,6-dimethoxy-9-(2-phenylethenyl) (RT:11.19 min), He ptasiloxane,1,1,3,3,5,5,7,7,9,9,11,11,13,13-tetradecamethyl6,8-Difluoro 2,2,4,4,6,7,7,8,9,9- decamethyl- [1,3,5,2,4,6,7,8,9] trioxahexasilonane (RT:14.28 min), 5-hydroxy-7-methoxyflavanone, tertbutyldimethyl-silyl ether (RT: 16.35 min), Hexadecanoic acid, ethyl ester (RT: 23.49 min), Linoleic acid ethyl ester (RT: 29.46 min), Ethyl Oleate (RT: 29.64 min) and Bis[di(trimethylsiloxy)-phenylsiloxy]-trimethyl-siloxyphenylsiloxane (RT: 29.75 min. and 34.9 min). *Ageratum conyzoides* (L.) L. extract (ACE) low antioxidant activity was observed.

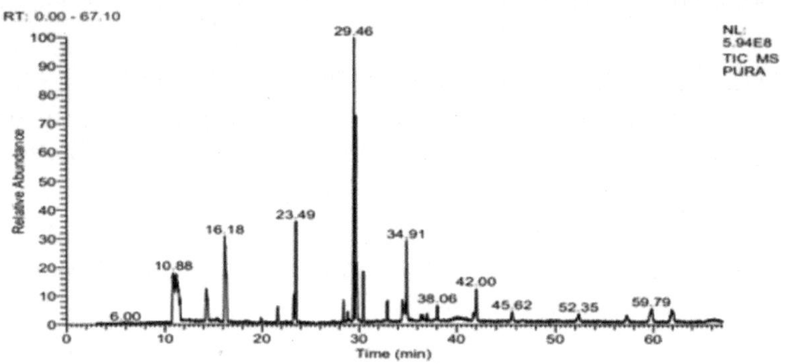

Fig: GC-MS Chromatogram of Ageratum conyzoides methanolic extract

CONCLUSION

The finding of the study concludes that out of 40 plant species under 19 families with highest percentage of family Asteraceae. These plants majority used were categorized under Agroforestry, medicinal, pharmacological, edible, ornamental, commercial etc. The second section of this study was Experimental analysis of *Ageratum conyzoides* (L.) L. The plant samples were collected and screened for phytochemical components. Phytochemical screening of the extract helped to identify the active constituents in the extract. Whole plants were used for study and extracted with ethanol. The screening results showed that *Ageratum conyzoides* (L.) L. showed the presence many anti-oxidants as glycosides, flavonoids, saponins, tannins, terpenoids and phenols and the absence of alkaloid and anthroquinone. Antioxidants have high value as they scavenge free radicals produced in the body cells and thus avoid or lessen the damage caused due to oxidation. In our study, antioxidant activity was tested using most common method of DPPH assay.

The objectives were defined with consideration that it would provide threshold informations on invasive plant species predominant in Gopalganj region. It would be basis of any future studies on invading species of this region. It could act as evidence of documented literature as this survey is one of its kind for this region. Documentation has been done for invasive alien species of Uttar Pradesh, Varanasi, Gorakhpur, Madhya Pradesh. There is only one study which accounts Invasive alien species of Bihar but not categorically for Gopalganj.

REFERENCES

1. Chatterjee, D (1940). Studies on the endemic flora of India and Burma. J. A McNeely, pp. 195-208. IUCN, Gland, Switzerland.
2. Dutta, S and Mukherjee, P (1916). Exclusive medicinal uses of alien invasive plant species in Dumka, Jharkhand. Conservation of Medicinal plant conventional and Modern approaches. pp.370-382.
3. Harborne, J.B. (1984). Phytochemical methods. 2nd edition. Chapman and Hall.
4. Junaid, S.A., Olabode, A.O., Onwuliri, F.C., Okwori, A.E.J and Agina, S.E. (2006). The antimicrobial properties of *Ocimum gratissimum* extracts on some selected bacterial gastrointestinal isolates. *African Journal of Biotechnology,* 5(22).
5. Levine, J.M., Vila`, M., DAntonio, C.M., Dukes, J.S., Grigulis, K and Lavorcl, S. (2003). Mechanisms underlying the impact of exotic plant invasions. *Proc. R. Soc. London, Serie B,* 270, pp.775-781.
6. Mooney, H.A., Hobbs, R.J. (2000). Invasive species in a changing World, Island Press, Washington, D.C.
7. Reddy, C and Sudhakar (2008). Catalogue of invasive alien flora of India. Life Science Journal. Vol.5. No.2.
8. Sofowora, A. (1993). Medicinal Plants and Traditional Medicines in Africa. *Chichester John, Willey & Sons New York.* 256.

9. Trease, G.E and Evans, W.C. (2002). Pharmacognosy. 15th Ed. London: Saunders Publishers. pp.42-44. 221-229, 246-249, 304-306, 331-332, 391-393.

10. Wall, M.E; Eddy, C.R; McClenna, M.L and Klump, M.E. (1952). Detection and estimation of steroid and sapogenins in plant tissue. *Analytical Chemistry,* 24, pp.1337-1342.

11

Bacterial Bioremediation of Endosulfan: Evaluating Psychological Impacts on Exposed Populations

Anupam Kumari and Deokant

Department of Botany, Samastipur College, L.N. Mithila University, Darbhanga (Bihar)

INTRODUCTION

The world's food demand is rising exponentially due to the expanding population. Many methods for growing crops efficiently have been adopted in order to increase agricultural productivity. Pesticides are chemical compounds used to eliminate pests. They are chemical or, biological agents, that weaken, incapacitate, and kill pests. Based on the types of targeted pests, the pesticides can be divided into several groups, namely insecticides, herbicides, rodenticides, bactericides, fungicides, and larvicides [Rafffaet *et al.,*2021].Pesticides, belonging to various chemical families are frequently employed for reducing crop yield losses caused by insects and vector-borne diseases, thus play a crucial role in global food production. Despite their significance, the use of fertilizers and pesticides raises environmental concerns due to the accumulation of substantial amounts in soil and water, negatively affecting the food chain and drinking water [Rahman *et al.,*2018]. The environmental hazard posed by pesticides stems from their enduring presence in the ecosystem, with an estimated annual spray of three million metric tonnes of insecticides worldwide. Many of these pesticides do not fully break down in the soil over time, leading to bioaccumulation and biomagnification when they enter the food chain. This phenomenon impacts both target and non-target organisms, including humans, and results in adverse effects on birds, wildlife, domestic animals, fish, and livestock.

Pesticide residues not only contribute to local environmental contamination of air, soil, and surface water but also present a major issue in food items and drinking water. Numerous published investigations have reported pesticide contamination in various products such as milk, honey, tea, etc. The health risks to humans depend on the type of pesticide and the degree of exposure, ranging from minor issues like headaches and skin rashes to more severe consequences such as neurological diseases. In rare cases, incorrect or misuse of pesticides can lead to paralysis, blindness, and even death. There have been marked psychological impacts of exposure to pesticides. Medina et al.,2019 conducted a study investigating the psychological impact of pesticide exposure on agricultural workers. The research, involving 140 individuals exposed to various pesticides, found a significant reduction in red blood AChE activity due to chronic exposure. This decrease in enzymatic activity was associated with the development of neuropsychiatric disorders, encompassing generalized anxiety and an elevated risk of suicide. Specifically, 25% of the participants met the criteria for a major depressive diagnosis accompanied by suicidal attitudes, while 23.9% exhibited generalized anxiety. Additionally, 23.5% displayed a combination of depression and anxiety, and 22% met the criteria for major depression [Medina *et al.*,2019]. A study among banana workers in Costa Rica, revealed a significant association between high-dose pesticide poisoning and elevated scores for anxiety, depressive symptoms, and personality disorders [Wesseling *et al.*,2002].

A recent systematic review in Environmental Research reveals an elevated risk of depression symptoms among pesticide applicators, supported by specific biomarkers. The study, following PRISMA guidelines, identifies a notable association between occupational pesticide exposure and depression symptoms, with 78% of reviewed articles supporting this link. The impact extends beyond occupational hazards, affecting mental health in agricultural workers, who face a six times greater risk of depressive symptoms. Pesticides, known to induce various health issues, are increasingly recognized as contributors to mental health concerns, with adolescents found to be at higher risk of depression.The scientific community has long entertained the idea that prolonged exposure to pesticides might elevate the likelihood of developing dementia. Moreover, it is theorized that chronic exposure to pesticides could impact distinct cognitive functions independently of a dementia diagnosis. Numerous studies have explored this potential correlation, emphasizing the differentiation between individuals with chronic exposure and those experiencing acute episodes of pesticide poisoning, a factor that may exert influence on the observed effects [Aloizou *et al.*,2020]. Due to the problems of pesticide contamination, for clean-up of pesticide contaminated soil, develop technologies that guarantee their elimination. Currently bioremediation is one of the most environmentally safe and cost effective method for decontamination and detoxification of pesticide contaminated environment. Endosulfan is a highly toxic and persistent organochlorine pesticide that has been widely used in agriculture to control various pests, including insects, mites, and nematodes [Yadav & Devi, 2017]. It was first introduced in the 1950s and gained popularity due to its effectiveness against a broad range of agricultural pests. it is chemically classified as a chlorinated hydrocarbon and is known for its long environmental half-life, which means it can persist in the environment for an extended period [Vivekanandhan &Duraisamy, 2012: Marcic *et al.*,2011]. Certain pesticides, such dieldrin and DDT [l,l,l-trichloro-2,2-bis-(p-chlorophenyl)ethane], have shown to be resistant. As a result, they are known to accumulate within food chains decades after being applied to soil and to stay in the environment for a very long period [Aberdeen,1980: Kannan *et al.*,1994].

According to many research, a large variety of microorganisms are capable of decomposing pesticides [Rani *et al.*,2008]. A diverse group of bacteria, including members of the genera *Alcaligenes, Flavobacterium, Pseudomonas* and *Rhodococcus,* metabolize pesticides [Aislabie & Lloyed-Jones, 1995]. Specific pesticides, including diazinon and carbofuran, biodegrade easily and can even prevent efficient pest control because of how quickly they break down in specific kinds of soil [Felsot,1989]. The isolation of indigenous bacteria capable of metabolising certain pesticides has received a lot of attention and is seen as an efficient method for pesticide bioremediation[Rahman *et al.*,2002]. The present investigation was aimed at isolating and identifying endosulfan-degrading bacteria from soil samples in the Samastipur district of Bihar.

MATERIAL AND METHODS

Collection of soil sample:

Soil samples from different farmlands near Samastipur (latitude 25.712219, longitude 85.804864) district of Bihar area was collected in duplicates. In each case one sample will be collected from soil contaminated with pesticides and another from pesticide free soil. After collection soil samples was collected in sample collecting jar and will be transferred aseptically to the laboratory.

Chemical and media:

Pesticide (Endosulfan) dissolved in acetonitrile to make a stock solution was filter sterilized and kept in refrigerator for use. Bushnell has agar media (p^H 7.0) was used for the isolation and cultivation of Endosulfan degrading bacterial strains.

Bushnell Haas (bh) medium composition:

$MgSO_4$	0.2g/l
$CaCl_2$	0.02g/l
KH_2PO_4	1g/l
K_2HPO_4	1g/l
$(NH_4)_2SO_4$	1g/l
$FeCl_3$	0.05g/l

Isolation of Coragen degrading bacteria:

Minimal media containing pesticide as a carbon source will be used to isolate pesticide (Endosulfan) degrading bacteria. Microorganisms are generally found in nature (air, soil and water) as mixed populations. The bacterial species were isolated from Endosulfan contaminated soil of Samastipur district (Bihar), by serial dilution method. Different bacterial colonies were observed in the Bushnell Haas agar medium. This colonies were identified by Gram's staining and biochemical tests. Pesticides will be used at different concentrations of 0.1%, 0.5% and 1% in Bushnell Haas Medium as a selective enrichment agent for the isolation of bacteria.

RESULTS

In this study, the following two bacterial genera were able to degrade Endosulfan: Flavobacterium and Bacillus.

Isolation of Coragen degrading bacteria:

The bacterial species were isolated from pesticide-contaminated soil of Samastipur district (Bihar), by serial dilution method. Different bacterial colonies were observed in the Bushnell Haas agar medium. This colonies were identified by Gram's staining and biochemical tests.

Endosulfan utilizing bacteria:

In this result of Endosulfan degradation of Flavobacterium and Bacillus. In *Flavobacterium,* the maximum zone of accumulation in 0.1%, 0.5%, and 1.0% at 2.17±0.40,1.93±0.3 and 0.77±0.20 respectively, and *Bacillus* maximum zone of accumulation is 0.1%, 0.5% and 1.0% at 2.28±0.50, 1.97±0.4, 0.87±0.10(Table 1).

Table.1 Total bacterial population and degradation of Coragen at concentrations of (0.01 and 0.1%) in contaminated soil.

S. No.	Organisms	Dilution	Total viable counts (CFU/g)	Endosulfan resistance bacterial counts (CFU/g)		
				Endosulfan concentration		
				0.1%	0.5%	1.0%
1	*Flavobacterium*	10^{-4}	5.50±0.30	3.55±0.40	2.54±0.3	1.00±0.20
		10^{-5}	4.25±0.50	2.54±1.40	2.33±0.2	0.70±0.10
		10^{-6}	3.76±0.45	1.41±0.30	1.34±0.6	0.65±0.55
		10^{-7}	3.54±0.45	1.30±0.20	1.39±0.3	0.56±0.55
2	*Bacillus*	10^{-4}	5.60±0.50	3.65±0.50	2.62±0.4	1.07±0.10
		10^{-5}	4.32±0.45	2.62±0.50	2.44±0.5	0.90±0.54
		10^{-6}	3.94±0.55	1.50±0.40	1.40±0.4	0.80±0.47
		10^{-7}	3.32±0.35	1.41±0.31	1.40±0.2	0.60±0.46

CONCLUSION

According to the study's findings, Flavobacterium and Bacillus are two particular bacteria that are capable of degrading endosulfan in soil. This discovery holds significance as it suggests the presence of endosulfan-degrading bacteria in the farmlands of Samastipur district, Bihar. This indicates the possibility of natural bioremediation of endosulfan-contaminated soils in this area. The process of neutralising or removing contaminants from contaminated environments by use of living organisms, including fungi, bacteria, or plants, is known as bioremediation. Therefore, the conclusion drawn from this study is that utilising Flavobacterium and Bacillus bacteria could be an effective strategy for bioremediation, specifically targeting endosulfan-contaminated soils in agricultural areas like Samastipur district, Bihar. These bacteria could potentially be harnessed in future applications or techniques for eco-friendly and sustainable management of pesticide contamination in agricultural soils.

REFERENCES

1. Aberdeen, I. (1993). Litigation over dieldrin in Victoria. Agricultural Science May, 42-5
2. Aislabie, J., and Lloyd-Jones, G. (1995). A review of bacterial-degradation of pesticides. *Soil Research*, 33(6): 925-942.
3. Aloizou, *et.al.,* (2020). Pesticides, cognitive functions and dementia: A review. *Toxicology Letters.* 31-51
4. Bentley, *et.al*, (2010). Woodward, Chlorantraniliprole: an insecticide of the anthranilicdiamide class. Hayes' Handbook of Pesticide Toxicology (Third Edition). Academic Press, NY 2010.
5. Felsot, A. S. (1989). Enhanced biodegradation of insecticides in soil: implications for agroecosystems. *Annual Review of Entomology*, 34: 453-76.
6. Kannan, *et. al.*, (1994). Persistent organochlorine residues in foodstuffs from Australia, Papua New Guinea and the Solomon Islands: contamination levels and dietary exposure. The Science of the Total Environment, 153: 29-49.
7. Medina, *et.al.*, (2019). Neuropsychiatric Disorders in Farmers Associated with Organophosphorus Pesticide Exposure in a Rural Village of Northwest México. *International Journal of Environmental Research and Public Health.* 16: 689.
8. Naqvi M and Vaishnavi C. (1993) Bioaccumulative potential and toxicity of endosulfan insecticide to non-target animals Camp. Biochem. Physiol. C. 105: 347-361 https://doi.org/10.1016/0742-8413(93)90071-R
9. Pandey *et. al.*, (2009). Biotransformation of the neonicotinoid insecticides imidacloprid and thiamethoxam by Pseudomonas sp. 1G Biochem. Biophys. Res. Commun. 380:710-714https://doi.org/10.1016/j.bbrc.2009.01.156

10. Raffa *et. al.*, (2021)., Remediation of Metal/Metalloid-Polluted Soils: A Short Review. *Applied Sciences.* 2021; 11(9):4134. https://doi.org/10.3390/app11094134

11. Rahman, *et. al.,* (2018). Isolation and Identification of Pesticides Degrading Bacteria from Farmland Soil. *Bangladesh Journal of Microbiology.* 35. 90-94. 10.3329/bjm.v35i2.42635.

12. Rani, *et. al.*, (2008). Isolation and characterisation of a chlorpyrifos degrading bacterium from agricultural soil and its growth response. *African Journal of Microbiology Research*, *2*(2): 26-31.

13. Wesseling, *et.al.*, (2002). Long-Term Neurobehavioral Effects of Mild Poisonings with Organophosphate and n-Methyl Carbamate Pesticides among Banana Workers. *International Journal of Occupational and Environmental Health*, 8: 27–34.

14. Marčić, D., Perić, P., & Milenković, S. (2011). Acaricides-biological profiles, effects and uses in modern crop protection. *Pesticides-formulations, effects, fate*, 39-62.

15. Vivekanandhan, N., & Duraisamy, A. (2012). Ecological Impact of Pesticides Principally Organochlorine Insecticide Endosulfan: A Review. *Universal Journal of Environmental Research & Technology*, 2(5).

16. Yadav, I. C., & Devi, N. L. (2017). Pesticides classification and its impact on humans and environment. *Environmental science and engineering*, *6*:140-158.

12

The Effects of a Combination of Five Medicinal Plant Extracts on the Growth and Sporulation of Pathogenic Fungi Prevalent in Makhana Plant

Sangeeta Kumari and Anand Mohan Mishra

University Dept. of Botany, Darbhanga, Bihar, India

INTRODUCTION

To manage the human health, there is importance of the healthy food. Now a days it is observed that the food which are in various modes, become not free from the access of chemicals because they are found to be chemically treated. Those chemicals loaded on the food materials are directly engulfed by the human beings through their food practices and these chemicals when enter inside the human body are toxic in nature and are responsible for creating a number of Physical abnormalities in the human body like Gastric disorders, Peptic Ulcer, Intestinal Cancer, lung disorder, Kidney damage, blood poisoing and the Heart failure. So, there is the need of producing the food stuffs free from the chemicals and finally producing the bold and bonafide health. Here there is question that how the food Stuffs become loaded with the harmful chemicals? The answer is that the standing crops become the platform of various pathogenic agents in the form of Bacterias, Viruses and Fungi together with also the different types of Insects and Pests. These pathogenic agents are responsible for causing the serious damage of the agricultural crops and finally into the agricultural production by causing the different diseases. To get off free from the ravages of these harmful pathogenic agents, the farmer apply different types of chemicals formulated for the destruction. of these harmful pathogenic agents and are available in the market in the form of Bactericides, Nematicides, Fungicides, Insecticides and Pesticides with different trade names. No doubt, the application of these chemicals control the growth and development of the pathogenic agents which are prevalent on the agricultural crops but they are stable and persistent which persist on the agricultural produce till the formation of grains and fruits. Some portions of it may enter inside the agricultural products and stay there. When they are harvested and sold in the market become loaded with these harmful chemicals. Now, there is need of search out the alternative of these chemicals from the herbal source. Most of the herbs contain the medicinal ingredients in the form of the Phenols, Terpenoids, Alkaloids, Flavonoids which according to Mahadevan 1982 are prohibitins which prohibits or inhibits the growth and sporulation of the harmful pathogenic fungi prevalent on the agricultural crops. These chemicals hidden inside the herbal source become harmless for the human beings who take the treated agricultural produce but harmful for the pathogenic fungi. It has been observed that these chemicals are temporarily existing on the agricultural plant that most of its part get evaporated in the environment and the residual may be eaten by bacterial agents. Makhana (*Euryale ferox salisbury)* is an important food stuffs for human beings which is rich Source of Protein and contain medicinal ingredients, Jha et.al-1991. They are generally produced in shallow and organic rich water. Naturally they are growing as healthy leaves but one of the harmful fungus known as *Alternaria* spp. cause the serious damage to the leaves by causing the harmful disease known as Leaf Blight. Due to the appearance of this disease, a large portion of photosynthetic area of the leaves get damaged resulting in the form of brown, burnt appearance. The product of Makhana usually in the form of Makhana Puff are sold in market and exported into the foreign countries. Before trading they are tested for the health certificate. If those Makhana puffs become loaded with harmful chemicals are out rightly rejected from the foreign market.

Considering the above views, I have decided to test a mixture of five medicinal plant on the health of four pathogenic fungi causing serious diseases on Makhana plant. A brief account of these medicinal plants are as such.

Neem (*Azadirachta indica*)

- *Azadirachta indica* commonly known as Neem, Margosa, Indian lilac, is a tree in the mahogany family Meliaceae. It is a medium-sized tree reaching 15 to 30 m in height with a large round crown 10-20 m in diameter. Neem has a deep tap root and is a mycorrhizal-dependent species.
- The leaves are alternate, Petiolated, clustered at the end of branches, unequally Pinnate, glabrous and dark glossy green at maturity, 20-40 cm in length and bearing 10-20 leaflets.
- The flowers are numerous fragnant white and borne in large clusters (up to 30 cm long).

- Neem fruits are 1-2 cm long drupes, smooth and green with white milky Juice when unripe, turning a yellow to brown when mature. They have a thin epicarp, a mucilagenous fleshy mesocarp and a hard endocarp.

They contain a variable number of ovoid (1-2 c.m.) oil seeds.

Chemical Constituent :

It is considered that *Azadirachta indica* shows therapeutic role due to rich source of antioxidant and other valuable active compound such as Azadirachtin (C35 H44 016), Nimbolinin, Nimbin, Nimbidin, Nimbidol, Epoxyazadiradione (C28 H34 06), Phytol, Gedunin, Salannin, Quercetin.

Medicinal Value :

Neem (*A. indica*) Plants parts shows antimicrobial role through inhibitory effect on microbial growth/ Potentiality of cell wall break down. Azadirachtin, a complex tetranortriterpenoid limonoid present in Seeds, is the key constituent responsible for both antifeedant und toxic effects in insects.

- Neem has proved effective against Certain fungi as fungicides. Neem preparations showed toxicity to culture of 14 common fungi including members of the following genera like Trichophyton, Ebidermophyton, Microsporum, Trichosporum, Geotrichum, candida.
- Pharmacological activities of Azadirachta indica in disease management through the modulation of various activities like Antidiabetic, Neuroprotective, Antiviral Antibacterial, erial, Antifungal, Antioxidant, Anti-inflammatory, imalarial Antimalarial, Heptoprotective, Antinephrotoxicity, wound healing effect, Immunodulatory effect, anticancer activity.

Tulsi (*Ocimum sanctum*)

Tulsi is classified as part of the Lamiaceae family, which includes flowering plants. *Ocimum* is a genus of aromatic herb and Shrub. Lamiaceae is a family of Plants that are native to tropical and subtropical regions. *Ocimum sanctum* (Tulsi) is an erect, much branched Sub-Shrub (30-60 cm tall) with simple opposite green or purple leaves that are strongly scented and hairy stems. Leaves have petiole arid are ovate, up to 5 cm long, usually somewhat toothed. Flowers are purplish in elongate racemes in close Whorls. It is cultivated for religious and medicinal purposes and for its essential oil.

Chemical Constituent :-

Ocimum sanctum is an aromatic Plant. The main Chemical constituent of Tulsi are: Oleanolic acid, Ursolic acid, Rosmarinic acid, Eugenol, carvacrol, Linalool, ß caryophyllene, germacrene. Volatile Oil, fixed oil also obtained from the leaves of medicinal herbs. Monoterpene are obtained from the volatile oils such as, camphene, myrcene, sabinene, in which some monoterpene produced oxygen such as linalool, borneol. Phytochemical analysis of this medicinal herb can identify the nature of compounds present in the extract of *Ocimum sanctum*. It is also for identify the bioactive compound and their effect.

Medicinal Value :-

1. Tulsi is cultivated for religious and traditional medicine purposes and also for its essential oil. It is widely used as herbal tea, commonly used in Ayurveda and has a place within the Vaishnava tradition of Hinduism, in which devotees perform worship Involving holy basil plants or leaves.
2. Tulsi is called the wonder herb or sometimes the holy herb because of its medicinal properties. Tulsi is also used to treat heart disease and fever. It is used to cure fever, common cold, sore throat, Kidney Stones, headaches and Asthma.
3. Tulsi might be an antipyretic agent, anti-inflammatory, antiemetic (Prevent vomiting), analgesic, anti-asthmatic, diaphoretic (induces sweating), anticancer, hebatoprotective agent, hypolipidemic activity (lower cholesterol) and antidiabetic also.
4. *Ocimum sanctum* considered as a potent adaptogen. Tulsi has a unique combination of Pharmacological actions that Promote well being and resilience.

Chrysopogon zizanioides **(Khus-Khus)**

Chrysopogon zizanioides, commonly known as vetiver and khus is a perennial bunch grass of family poaceae.

Vetivers widely cultivated in tropical regions of the world. It has thick fibrous adventitious roots. Which are aromatic and highly valued. Unlike most grasses it forms horizontally spreading mat like root systems. Khus roots plant are long, thin and fibrous and can reach up to 4 meters in depth. The stems are tall and the leaves are long, thin and rather rigid. Leaves are large and tufted with bunchy grasses. It grows upto the height of 5 feet. Inflorescence are spikelets and there are three stamens. The flowers are brownish purple. Khus plants have a gregarious habit of growing and they grow in clusters.

Chemical Composition :-

The roots of this grass on steam distillation yield an essential oil mainly consisting of sesquiterpenes (3-4 %), sesquterpenols (18-25 %) and sesquiterpenones (7-8 %). Among these the major economically important active compounds are khusimol, a- vetivone, b- vetivone which constitute about 35% of oil.

Medicinal Importance:

Khus-khus has been found effective in treating insomnia. The oil extracted form khus-khus relaxes and soothes the tired muscles and the mind. It help in reducing inflammation.

Calotropis procera **(Arka)**

Calotropis procera is an evergreen perennial shrub of the family Asclepiadaceae, mainly found in arid and semi-arid region. It is a large shrub growing to 4 m (13 ft) tall. The branch bears opposite leaves which are simple, petiolate, ovate with entire margin and pointed tip. The young leaves are covered by small, soft, valvatae hairs and bears white milky latex. The older part of the stem is hard, smooth and covered by bark. The younger stem is green, shiny, soft and glaborous which also bears the white milky latex. Leaves with acute subsessile apices extraaxillary umbellate panicle infloressence with purple corolla and erect lobes. It has clusters of waxy flowers that are either white or lavender in colour. Each flower consist of 5 pointed petals and a small "crown" rising from the cener which holds the stamens. The aestivation found in calotropis is valvate.

Calotropis is a large bushy shrub with decussate, obovate, coriaceous, auriculate, leaves with acute subsessile apices extraaxillary umbellate panicale infloressence with purple corolla and erect lobes.

Chemical Constituent :-

Calotropis procera contained many biological active chemical group including cardenolides, steroids, Tannins, glycosides, phenols, Terpenoids, sugar, Flavonoids, Alkaloids and saponins. Latex produced by C. procera contains many compounds like alkaloids. They are calotropin, catotoxin, calcilin and gigantean. These alkaloids are very harmful and having poisonous properties.

Medicinal Value :-

Calotropis contains chemicals that might help thin mucuous and make it easier to cough up. *Calotropis* has shown some activity against pain, inflammation, bacteria, fever and ulcers caused by alcohol and medications such as aspirin, indomethacin (indocin) and others. The leaves of c. procera are said to be valuable as an antidote for snake bite, sinus fistula, rheumatism, mumps, burn injurics and body pain. The leaves are also used to treat jaundice.

Cascabela thevetia (kaner)

Cascabela thevatia is an evergreen tropical shrub belong to family *Apocynaceae*. It is a poisonous plant native throughout Mexico and in Central America and cultivated widely as an ornamental. It is effectively drought resistant and tolerant to high temperature, hence found in various state of India. It is commonly known as kaner. It is a relative of Nerum oleander, giving it is a common name yellow oleander.

C. thevetia is an evergreen shrub with thick and pointy leaves arranged in pairs or whorls of 3 along the stem. This fast growing tropical shrub produces clusters of reddish 5-petalled flowers at the end of branches and are sweetly scented. Its stem is green turning silver/grey as it ages. Its leaves are willow-like, linear-lanceolate and glossy green in colour. They are coverd in a waxy coating to reduce water loss. The long funnel-shaped some times fragrant yellow (less commonly apricot, sometimes white) flowers are in few-flowered terminal clusters. Inflorescence a terminal cyme. Its fruit is deep red black in colour encasing a large seed that bears some resemblance to a "chinese lucky nut".

Chemical Constituents :-

Thevetin A and Thevetin B are the main cardiac glycosides found in yellow oleander seed. In addition to Thevetin A and B, Peruvoside, Neriiflin, thevetoxin, Peruvosidic acid, Cerberin, thevetoxin, ruvoside, acetylneriifolin and thevefolin are some of the other reported glycosides, yellow oleander reported to consist of 2 types of sugar- D glucose and L-thevetose. Various aglycones are also found in

C. thevetia are :- Cannogenin, Cannogenol, Cannogenic acid, Digitoxigenin.

Medicinal Value :-

C. thevetia is used as Biological pest control. The plant's toxin have tested in experiments for uses in biological pest control. Its seed oil was used to make a "paint" with antifungal, antibacterial and anti termite properties.

- All parts of the *C. thevetia* plants are toxic to most vertebrates as they contain cardiac glycosides. The main toxin are the cardenolides called thevetin A and thevetin B. Other include peruvoside, nerifolin, thevetoxin and ruvoside. These cardenolides are not destroyed by drying or heating. They produce gastric and cardiotoxic effect.
- Antidotes for treatment include atropine and digoxin immune fabs (antibodies) and treatment may include oral administration of activated charcoal.
- Kaner is used in making madicines for leprosy, inflammation, wound, heart etc. Kaner whole plant is said to have anticancer properties, help in leprosy, malaria, ringworn, indigestion.

PREPARATION OF EXTRACT IN COLD DISTILLED WATER

At first 250 gm of fresh leaves of each early described medicinal plants were separately taken and washed thoroughly with tap water and finally with distilled water to remove the outer unwanted material fixed on the upper parts of the leaves. Now they are dried in hot oven or incubator at 80°C for 2 days and the dried leaves of each medicinal plants were taken separately in a mortor and were crushed with pestals. The crushed material now were taken separately in a mixer grinder to get fine powder and the powdered material were collected separately. Later on 2 grams of the powdered material of each plant were finally subjected to prepare the aqueous solution of 2 percent. These aqueous solution were taken separately in a bolted tubes and were kept in 5000 rpm centrifuge. Now they were rotated for 5 minutes and allowed the liquid material to settle down. The pure liquid settle at the bottom of the tube and erhanious matter come at the top of the liquid. They were removed and the pure extract was obtained. The pure extracts of all the five plants were allowed to settle completely for one day and after that a mixture of extract is prepared as such.

Neem extract – 50 ml

Thevetia extract – 5 ml

Tulsi extract – 10 ml

Calotropis extract – 7 ml

Khus extract – 5 ml

These combination was processed in a separate conical flask of 25 ml and was shaken well. They were bolted tightly and was kept safely for antifungal bioassay.

EXPERIMENTAL FINDING

The experiment is carried out considering four pathogenic fungi which become prevalent on Makhana plant like *Alternaria* Spp., *Fusarium solani*, *Rhizoctonia fabi* and *Sclerotinia sclerotiorum*. All these four fungi were grown separately on seven days old culture of PDA media and were transferred to five different petri plates containing antifungal extracts. After seven days the result were recorded in diameter and sporulation were counted. All results were placed in table as such.

Table 1:- Result showing the growth and sporulation of fungi at different concentration of plant extract :-

Sr. No	**Concentration**	**Name of Pathogens**							
		***Alternaria* Spp.**		***Fusarium solani***		***Rhizoctonia fabi***		**Sclerotinia sclerotiorum**	
		Mycelium diameter	Sporulation	Mycelium diameter	Sporulation	Mycelium diameter	Sporulation	Mycelium diameter	Sporul.
1	5 ml.	5 mm	++	15 mm	+++	8.1 mm	++	10.6 mm	++
2	10 ml.	2 mm	+	10.5 mm	++	6.3 mm	++	8.4 mm	++
3	15 ml.	--	--	8.3 mm	+	4.1 mm	+	6.3 mm	+
4	Control	40 mm	++++	36.3 mm	++++	38.3 mm	++++	46.1 mm	++++

Cd - 5 % – 0.1238

SE - 1 % – 0.367

+ = Poor Sporulation

++ = Medium Sporulation

++++ = Excellent Sporulation

- - = No growth and Sporulation

RESULTS AND DISCUSSION

From the above table it is evident that almost all medicinal plants contained the antifungal ingredients. They are successfully controlling the growth and sporulation of pathogenic fungi. In this result, it has been observed that the extract shows potent antifungal activities in case of some pathogenic fungi and in case of other the effect is moderate. The *Alternaria* spp. Shows more inhibitions in comparison to *Fusarium* and *Sclerotinia*. However it has more powerful activities in case of *Rhizoctonia fabi*.

From perusal of table 1, it is evident that the aqueous extract of *Thevetia* leaf has potent antifungal constituents which completely seizes the growth of the fungus from both the aspects i.e. the mycelial and conidial. This finding is in accordance with the finding of Shrivastava and Lal (1997). They found the leaf extract of *Calotropis procera, Lantena camara, Ocimum basilicum* and *Azadirachta indica* stopped the mycelia growth of *Curvularia tuberculata* and *Alternaria alternata.* They also reported the maximum toxicity against spore germination. Further this result justifies the result of Bansal and Gupta (2000) who evaluated some plant extract against *Fusarium oxysporum* a wilt pathogen of Fenugreek. They have reported the inhibition of growth and sporulation by the extract of *Calotropis procera, Azadirachta indica, Atropa belladonna* and *Lantena camara.* This work is in accordance with the findings of Zewain et al, (2004) who studied the effect of fungicides and neem extract on the mycelia growth of *Sclerotinia sclerotiorum* and reported the mycelia growth only 2 mm in diameter. In this work the mycelia diameter of Alternaria has been recorded 2 mm at 5 ml dose.

CONCLUSION

From the present work, it can be concluded that the nature has provided the mankind a lot of facilities for their protection but there is need of proper search and investigation. Higher plants grow in nature and contain a lot of chemical constituents which can be applied to treat the infected plants. These chemicals work like commercial fungicides. As we know that these commercial chemicals by nomeans are useful for the health of mankind but still we are applying these on our infected crops. Finally we take these chemicals with the agricultural produce and get result in the form of different health disorders. If such type of research work is carried out and ecofriendly chemicals will be found out then it may be possible that our infected crops will be treated with the beneficial chemical which at any dose will be no harmful for our health. In this way we will be able in providing some beneficial aspects to the suffering society.

After viewing the above result which is presented in table, it is concluded that, the extract prepared from the five medicinal plant could be able to control the spread of pathogenic fungi on Makhana plant and it will be proved the alternatives of the chemical fungicides formulated for the control of pathogenic fungi. The prophylactic spray of the extract on Makhana leaves will be helpful for control the *Alternaria* Leaf Blight and *Rhizoctonia* Stem decay. If it is possible, then why there should be the use of chemical fungicides which become harmful for the health of human beings and further if the Makhana product may be found to be loaded with this harmful chemical gives bad impact on the foreign trade market. The antifungal constituent found inside the herbal source has no side effects and there greater mass becomes not harmful at any concentration. So it is suggested that the Government should promote the herbal industry and give priority to obtain the antifungal constituent from herbal source. There are two benefits from this are following:

(1) Bonafide health of human being.

(2) Low cost of this chemical will help the poor farmer.

REFERENCES

1. Bansal, R.K. and Gupta, R.K. (2000). Evaluation of plant extracts against *Fusarium oxysporum*, the wilt pathogen of Fenugreek, IPS, Vol.53(1):p.107.108.
2. Bansod, S. and Rai, M. (2008). Antifungal Activity of Essential Oils from Indian Medicinal Plants against Human Pathogenic Aspergillus fumigates and A.niger, World,j.Med.sc.,3(2):81-88
3. Bhatt, Meenakshi, Mishra, M., Singh, J.V. and Anita. (2010). Phytochemical and Antimicrobial studies of *Adhatoda zeylanica*, IJRAP, 1(1),90-97.
4. Gerard Ezhilan,J. Chandrasekhar, V. and Kuruchev,v. 1994. effect of Six selected plant products and oil cakes on the sclerotial production and germination of *Rhizoctonia Solani* Ind. Phytopath .47:183-85.
5. Jha,V:Barat, G.K. and Jha, U.N (1991). Nutritional Evaluation of *Euryale ferox salisb.*(Makhana). Journal of Food science and Technology, 28(5):326-328.
6. Khan, J.A. and Tewari, S. (2011). A study on Antibacterial Properties of *Rosa indica* against various pathogens, Asian,j.plant sc. And Res.,1(1):22-30.
7. Mahadevan, A.(1970). Prohibitin and disease resistance. Phytopath.Z. Vol.68:p.73-80.
8. Antifungal potentiality of some medicinal plant extracts against Bipolari *Bipolaris Oryzae*. Asian Journal of Plant Science and Research, 1(3):77-80.
9. Masood, N., Ahmed Chaudhry, A. and Perween, Tariq (2008). In vitro Antibacterial Activities of Kalonji, Cumin and Poppy seed, Pak,j. Bot.,40(1), 461-467.
10. Shrivastava, A.K. and Lal, B.(1997). Studies on biofungicidal propertie of leaf extract of some plants. Ind. Phytopath.,50(3):408-411.
11. Zewain,Q. K., Bahadur, P., Sharma,P.(2004). Effect of fungicides and Neem extract on mycelia growth and mycelio genic germination on *Sclerotinia sclerotiorum*. Ind. Phytopath. 57(1):101-103.

13

Edible Wetland Plants Use as Medicine of Local People in Madhubani District, North Bihar

Ajay Kumar [1] and Mustafa Kamal Ansari [2]

[1]*Dept. of Botany, Parwati Science College, Madhepura*
[2]*Millat College, Darbhanga (L.N.Mithila University Darbhanga)*

INTRODUCTION

The flood plains of the Madhubani need to be explored to harness the potentials of several other aquaphytes as fodder, food, medicine and other livelihood purposes Kumar (2018). The edible wetland plants provide staple food indigenous people which serve as complementary food for other local inhabitants and an alternative source of cash through income the sale in local markets. These are important nutrient and vitamin supplements for indigenous people and contribute to reduce the vulnerability of local communities to food insecurity by providing a buffer in times of food storage. Local and rural people identify and collect these plants for food from wetlands and developed a range of processing methods in accordance with their needs. With modernization and setted agriculture this knowledge is becoming lost a trend that may lead to decreased diversity of indigenous diet and poorer nutrition Jha *et al.*, 2014, Kumar *et al.*, (2016,2017,2019). The aquatic and semi-aquatic plants of Madhubani district are rich repositories of various plant species such as *Acorus calamus, Alternanthera philoxeroides, Bacopa monnieri, Centella asiatica, Ceratophyllum demersum, Cyperus iria, Cyperus rotundus, Eclipta prostrate, Euryale ferox, Hygrophila auriculata, Ipomoea aquatica, Nelumbo nucifera, Nymphaea sp., Nymphaea nouchali, Polygonum glabrumWild, Polygonum hydropiper, Ranunculus sceleratus, Trapa bispinosa, Vetiveria zizanioides.*

MATERIAL AND METHODS

A survey to collect the detail of some edible wetland plants in Madhubani district was undertaken during March 2017 - February 2019. We recorded different group of macrophytic plants such as marginal, submerged, floating and emergent. Herbarium of each plant species was prepared using standard herbarium paper and techniques and the same have been deposited in the departmental herbaria plants were identified with the help of available literature of Subramanyam (1962), Cook (1996) and Choudhary (2002). The identification were also confirmed by comparing the herbarium specimens with voucher specimen kept in herbarium of university department of Botany, L.N.Mithila University, Darbhanga.

Table-1 Major Edible wetland plant use in food and medicinal purpose of Madhubani district.

Sl. No.	Botanical Name	Family	Local Name	Parts with medicinal properties or uses
01	*Acorus calamus*	Araceae	Achhaini/ Bach	Dried rhizome is used in Diarrhoea, Dysentery, Bronchial Catarrh
02	*Alternanthera philoxeroides*	Amaranthaceae	Sarhanchi	Plant is used in diarrhoea and dysentery.
03	*Bacopa monnieri*	Scrophulariaceae	Brahmi	Plant is antioxidant, aperient, aphrodisiac, astringent, cooling, diuretic, laxative and tonic
04	*Centella asiatica*	Apiaceae	Brahmi	Leaf powder is mixed in milk and consumed to cure madness.
05	*Ceratophyllum demersum*	Ceratophyllaceae	Kacher	Plant is antiperiodic, antipyretic, astringent, cooling and laxative
06	*Cyperus iria*	Cyperaceae	Motha	Plant is astringent, stimulant, stomachic and tonic
07	*Cyperus rotundus*	Cyperaceae	Chichorh/ Mothi	Tuber is eaten in flatulence. rhizome is starch rich with vitamins
08	*Eclipta prostrate*	Asteraceae	Bhangraiya	Plant juice is taken in jaundice. Leaf paste is used as hair tonic.
09	*Euryale ferox*	Nymphaeaceae	Makhana	Perisperm is consumed as tonic and astringent, Dessert delicacy, Effective against deficiency disease like Beri-Beri.
10	*Hygrophila auriculata*	Acanthaceae	Gokhura/ Talmakhana	Root decoction is used in the treatment of jaundice. Leaves are used as greens to increase the level of haemoglobin in blood.
11	*Ipomoea aquatica*	Convolvulaceae	Karmi	Young leaves are cooked and eaten in gastric and intestinal disorders.

Sl. No.	Botanical Name	Family	Local Name	Parts with medicinal properties or uses
12	*Nelumbo nucifera*	Nymphaeaceae	Kamal	Rhizome powder is taken to cure piles.
13	*Nymphaea sp.*	Nymphaeaceae	Sorkha	Rhizome paste is applied as ointment on infected portion and taken along with seed paste of black pepper (*P. nigrum* L.) as antibiotic in carbuncle.
14	*Nymphaea nouchali*	Nymphaeaceae	Bhent/ Malkoka/ kumudini	Rhizome powder is taken in indigestion, dysentery,piles etc. Rhizome/flower is mixed with seeds of black pepper(*Piper nigrum* L.), pasted and consumed in menstrual disorder;flower is more effective than rhizome in the rapeutic action.
15	*Polygonum glabrum*	Polygonaceae	Tota	Plant is antiviral and hypotensive. Root paste is applied on scabies.
16	*Polygonum hydropiper*	Polygonaceae	Panimirich	Leaves are diuretic, stimulant, styptic and lithontriptic
17	*Ranunculus sceleratus*	Ranunculaceae	Jaldhania	Leaf paste is applied on the body portions infected by skin diseases.
18	*Trapa bispinosa*	Trapaceae	Singhara	Cotyledons are consumed to stimulate the desire to eat.
19	*Vetiveria zizanioides*	Poaceae	Katra grass	Its oil provides relief from rheumatism, headache cosmetic industries
20	*Schoenoplectus articulates*	Cyperaceae	Khobhi	Pops of seeds are given to the pox patients to speak properly.

RESULTS AND DISCUSSION

Each and every mentioned plant is useful in some way or other because they are untapped reservoirs of drug. The Structural diversity of their chemical components makes Them a valuable source of drugs that control several metabolic disorders and microbial infections. In recent time there has been a market shift towards herbal cures because of the pronounced cumulative and irreversible reaction of modern synthetic drugs. Aquaphytes like karmisaag (*Ipomoea aquatica*), kamal (*Nelumbo nucifera*), Bach (*Acorus calamus*) etc. have been found to be in possession of hypoglycaemia propertics which are also used as subsidiary food by the indigenous population.

Bach-dried rhizomes are possess antibacterial, anathemic properties and also used for treatment of chronic diarrhoea dysentery, bronchial catarrh and tumors. Brahmi (*centella album*) having neurotinic properties. This herbal plant has ascorbic acid as its natural ingredient can through to destroy scavenging free radicals and less on oxidative stress to considerable extent. Kamal (*Nelumbo nucifera*) fruits and seeds are used against diarrhoea, Bleeding, fever and its dried root, rhizome are used to treat gastrointestinal, genital and bronchial condition. *Alternanthera philoxeroides* is sold in the market as green vegetable and also used as medicinal purposes against acute cough and intestinal worm. Vetiver plant use described in Ayurvedic system and its oil provides relief from rheumatism Headache on local application.

The root part of the plant is now being used as refreshing drink. Eating row singhara (*Trapa bispinosa*) contributes to the ingestion of the giant intestine flude larve and also having high nutritious.Nymphaeaea spp fruit used against dyspepsida and dysentery. Makhana (*Euryale ferox*) serves as a dessert delicacy and is effective against deficiency disease like Beri-Beri. It act as cardiac stimulant and has binding action in dysentery. It is known in the East and South East Asia for its astringent, deobstruent, antirheumatic, anti diuretic and several other medicinal properties Jha *et al.*, 2014, 2018, Kumar *et al.*, 2022).

CONCLUSION

Wetlands have played a dominant role in shaping the culture, socio-economic needs and civilization of a particular place. Wild wetland plants play pivotal roles in their daily lives particularly in supplying dietary food items. It also provide livehood to a substantial section of society living around them. Hence, proper documentation of traditional knowledge related to these plants and identification of potential species for prioritization of conservation through sustainable management is essential for socio-economic development of the rural people of this area are great connoisseurs of rich biodiversity. Suggestions may be made to Bihar government for renaming the above act as Wetland protection act which should ensure that No one can encroach wetland areas for urban extension, Agriculture and other purpose.

REFERENCES

1. Choudhary, M.K., (2002). Aquatic Macrophytes of Jajibabad (Bijnor) U.P., J. Indian bot. soc 81. 11-15.
2. Cook, C., (1996). Aquatic and wetland plants of India, Oxford University Press, p 385.

3. Jha, A.K., Singh, C.B., (2018). Diversity and Perspective of Medicinal macrophytes in the wetlands of North Bihar, India, flora and fauna 24(1) 11-15.

4. Jha, V., Verma A.B., Kumar, R and Jha, M.S., (2014). Ethnic and economic aspects of some emergent aquaphytes in Mithila region of north Bihar. Ethnobotany 26 : 10-15

5. Kumar, A., (2018). Ecological management of wetlands of Madhubani with special reference to wetland resources. Ph.D. Thesis, L.N. Mithila University, Darbhanga, 236p.

6. Kumar, A., Bazmi, S.H, and shashi, S.B., (2018). Commercial value of aquaphytes growing in Rahika Block of Madhubani District in Bihar, India, Genius 6 (II): 7-15.

7. Kumar, A., Bazmi, S.H. and Jha, N.K., (2017). Nutritional and cultural use of Makhana (Euryale feroxSalisb) in north Bihar, India. Bioglobia 4 (2) : 42-46.

8. Kumar, A., Bazmi, S.H. and Rahmatullah, (2016). Biodiversity and their socio-economic importance of wetlands of Madhubani district (North Bihar), Bioglobia 3(1) : 28-34

9. Kumar, A., Bazmi, S.H. and Rahmatullah, (2017). Wetland management and utilization of North Bihar with special reference to wetland resources., Wisdom Herald 8(3):323-328.

10. Kumar, A., Kumar, S. (2019). A Case Study of Edible Wetland Plants and their contribution to livelihood of local people in Madhubani District, North Bihar. Ajanta 8(2) : 45-50.

11. Subramanyam, K., (1962). Aquatic Angiosperms, C.S.I.R., New Delhi.

12. Thakur, L.K., Choudhary, B. and Thakur., N.K., (1995). Aquacrops of Darbhanga district in north Bihar and their commercial significance. Journal of the Indian Fisheries Association (Proc. Nat. Symp. Aqua crops) p 107-111.

14

Emergence of Makhana (*Euryale ferox* Salisb.) as a Superfood during Post-Covid Dispensation

Chandra Bhushan Kumar Sinha[1] and Vidyanath Jha[2]

[1]*Dept. of Biotechnology, L.N. Mithila University, Darbhanga-846004, Bihar, India*
[2]*L.N. Mithila University, Darbhanga-846004, Bihar, India*

INTRODUCTION

Makhana, botanically known as *Euryale ferox* Salisb., belongs to the family Euryalaceae, carved out of the big family Nymphaeaceae, famous as the water lily family. The plant is known for producing its nutritious seeds that are mostly consumed in its popped form. The intricate job of its cultivation, harvest and post-harvest operations are generally performed by the fishing community of the Mithila region in Bihar and the adjoining district of bordering Nepal and Bangladesh. About 10 districts of North Bihar- Madhubani, Darbhanga, Supaul, Madhepura, Saharsa, Purnia, Katihar, Araria and Kishanganj are known as the areas where it is cultivated. 80 to 85% of the total makhana production in India comes from these districts.Makhana is known for its best quality of protein and starch . It is a much sought after food item in countries where more people suffer from obesity. It is an account of the fact that it is almost fatless. Makhana is held as a significant nutraceutical on account of its being rich in nutritional contents as well as acting as a remedy for a number of human ailments. It drew a significant global attention during the COVID pandemic. With the setting up of National Research Center at Darbhanga, a national mandate has been set for raising its production as well as bringing more and more of waterlogged areas in the country under its cultivation. Three Makhana Mahotsavs held at Patna in recent years bear testimony to this fact. Makhana stalls have now become a normal feature in trade fairs at national level.

Makhana as a significant nutraceutical

The edible part of *E. ferox* seed is the perispem. Raw seed is converted into popped form through an arduous method of post-harvest processing. Makhana seed with moderate 10-12 percent protein content is known for its high essential amino acid index (EAAI) of about 90 percent. This makes it comparable to fish and mutton so far as the quality of protein is concerned. Raw seed powder is an essential ingredient of the baby foods in China with a dependable system of indigenous medicine. It has about 78 percent carbohydrate mainly (**Table 1**) in the form of starch of very minute dimension (i.e., 1 to 3 micron). Makhana pops have a good prospect of being exported to affluent countries where there is a large population of obese people who are in search of non-fat diets. It has only 0.1% fat content and is rich in minerals (Zhang *et al* 2011). Properties of *E. ferox* starch have also been investigated (Liu *et al* 2006, Zhiwei *et al* 2011).

Table 1. Nutritional properties of Makhana.

Parameters	Content
Seed kernel	
Protein	9.70%
Carbohydrate	79%
Fat	0.10%
Iron	1.4mg/100g
Calorific Value	
Raw seed kernel	362kcal/100g
Popped form	328kcal/100g
Essential Amino Acid Index	
Raw seed kernel	93.63%
Popped form	89.95%
Biological Value	
Popped form	55
Chemical Score (% Egg)	
Raw seed kernel	70%
Popped form	56.57%
Comparative values of (a). Leucine/Isoleucine	
Raw seed kernel	1.84
Popped form	1.99
(b). (Arginine+Lysine)/Proline	
Raw seed kernel	6.3
Popped form	4.74

The raw seed of Makhana possesses a calorific value of 362 Kcal/100gm as against 328Kcal/100gm in its popped form. However, its biological value (B.V.) is low (around 55) which may be attributed to the high ratio of leucine to isoleucine amino acids in its seeds. Wild populations of "Thangjing" (Manipuri vernacular of Makhana) are also available in Manipur lakes. Nath and Chakraborty (1985a.b) have investigated protein-quality and starch in the Tripura population of *E. ferox*. Tribals in Tripura use the aril of *E. ferox* over its raw seeds (embedded inside fruits) for the cure of gonorrhea. Meitei tribals in the Manipur valley utilize several non-seed parts including young petiole, leaf, fruit coat, placenta and also the young seeds for preparing vegetable dishes. No such practice of utilization of vegetative parts, however, exists in north Bihar. Ayurveda mentions it as spermatogenic and aphrodisiac. In Mithila (India), it is customary for a new bride's parents to send Makhana to the grooms on the occasion of Kojagara festival (a marital ritual) (Jha *et al* 1991a,1991b). It is an integral component of a number of ethnic practices in Mithila and other parts of India (Kumari *et al* 2017).There is a need to adopt this practice in other Makhana growing areas of the country as Makhana ponds do have an excessive number of young plantlets that need to be thinned out to facilitate the maximum expansion of only the required number of plants. The giant leaves occupy a large space. One single *E. ferox* plant, if allowed unhindered space, may develop as many as eight leaves at a time, including expanding, maturing and senescing leaves (Kumari *et al* 2013, Singh *et al* 2003,2006; Jain *et al* 2010). Others have elaborated the cultivation and conservation practice as prevalent in Manipur. *E. ferox* growing in the north eastern state of Tripura has attracted scientific attention. A recent analysis (Bhowmik *et al* 2012) has found this aril to be rich in micronutrients. They have also reported the ratio of K/Na in *E. ferox* aril to be 7.19 and this plays an important role in the prevention of hypertension and atherosclerosis.

Table 2. Nutraceutical formulations Made of Makhana

Name	Country	References
Nutri-bar	India	[Jan *et al* 2012]
Yishen Capsule	China	[Fang 2003, Fang *et al* 2003, Fang *et al* 2005, Yu *et al* 2009, Fang *et al* 2013,]
Herbal Suppository	Korea	[Tsai and Cai 2014]

Growing market potentiality of Makhana is apparent from the act that its price during the year 2024 rose to as high as Rs. 1400 per kg. for its pop in Darbhanga market and Rs. 300 per kg for its guris (raw seeds). Their prices during the year 2019, five years back, were Rs.500 and Rs.120 respectively.(Table 3).

Table:- 3 Comparison of the rate of increase in price of Makhana seed and pop in recent years

Types	Current rate (2024)	Previous rate (2019)
Raw seed per kg	Rs. 300	Rs. 120
Popped Makhana per kg	Rs. 1200	Rs.500

Best quality pops are usually beyond the reach of the common man even in the area where it is grown. The crop often suffers from a dearth of water during the peak period of growth. This leads to the seeds remaining immature causing the formation of low-grade pops. Good quality Makhana usually called Rasgulla (or lava), constitutes roughly 54 percent of pops, followed by 35 and 11 percent, respectively, of the medium quality Murra and the low-grade Thurri types. Pops are used as snacks: sweetened form (called Makhana paagal), kheer, halwa, dal makhni, kofta, dum aloo, fried form (mostly used by those who perform the ritual fasts, by frying Makhana in ghee and adding rock salt to the same) (Kumari *et al* 2013).

As per market preferences Makhana is being used for making Panjiri, Idli, Dosa, Kulfi, Rayata, Burfi, beverages and a host of other culinary items, often in combination with millets that are being prospected in India for quite sometime. Investigations are also being made on developing full proof Makhana packagings (Sharma *et al* 2024). Some other culinary items of Makhana include Palak Makhana, Makhana curry, Makhana pulao, Nutty Makhana curry, Matar Phool Makhana, Choco Makhana, Makhana Chops etc. The traditional Ayurvedic system of Indian medicine holds Makhana as beneficial in '*Vata*' and '*Pitta*' disorders. It acts as an expectorant, emetic and as a cardiac stimulant. Both seeds and flowers have aphrodisiac properties. It forms a component of the Chinese medicinal formula 'Su-shin', a tonic especially required for the growth of children (Hsu *et al* 1951). The addition of popped Makhana flour improves the quality of the bun and cookies (kumar *et al* 2014,2015). Makhana flour is utilized in combination with the flour of another water fruit called chestnut (*Trapa* sp.) to prepare an edible product called Pua (Shamim *et al* 2017). A ready-to-mix infant food using a gorgon nut, samak rice and banana powder has been formulated [Bana *et al* 2015]. Makhana is a significant nutraceutical and is a good source of micronutrients (Jha *et al* 2016). Zhao et al (1989) attributed the medicinal properties in this plant to the presence of glucosyl sterols as active components (Zhao *et al* 1989). Anti-oxidant activity of plant products has been studied in correlation to its therapeutic properties including a reduction in ischemic/reperfusion cardiac injury [Lee at al 2002, Das *et al* 2006]. Two new cerebrosides and two new tocopherol trimers from the seed of this plant have been reported recently (Rao *et al* 2007,Zhao *et al* 2007). Its extract has also been demonstrated as a potential chemotherapeutic agent against neuronal diseases (Lee *et al* 2009). A recent study shows E. ferox as a proteinuria inhibitor of diabetic nephropathy (Song *et al* 2011). Its seed extracts have been examined as potent scavengers as well as inhibitors of LDL oxidation which is attributed to ischemic heart problems (Kim *et al* 2011). Closely agreeing to this observation, extracts from seeds of this aquaphyte have been demonstrated as significantly hypoglycemic as well as hypolipidemic [Ahmed et al 2015]. The antioxidant effects of the extracts have been attributed to these effects in rats. However, the role of flavonoids, polyphenols and alkaloids in this plant may not be ruled out.

E. ferox is used in the Korean traditional medicine for the control of dysentery, diarrhoea, leucorrhoea, incontinence and paralysis of joints. Makhana seeds are also understood to possess immunoral properties (Puri *et al* 2000). A number of other global studies have mentioned several medicinal aspects of this plant (Guo *et al* 1993,Li *et al* 2007,Song *et al* 2009,Zhang *et al* 2009). These studies suggest that this aquaphyte has good prospects of preventing cardiocerebral diseases, therefore, a good source material for developing new drugs and health products including flavonoids, cyclic peptide, sterol, brain lipid etc (Shen *et al* 2012). More recently, the effectiveness of aquatic macrophytes including that of *E. ferox* for the control of diabetes has been reviewed [Jha *et al* 2014]. A new compound 2β –hydroxy betulinic acid 3β caprylate (HBAC) isolated from E. ferox seeds has been found to be effective in glycemic control in streptozotocin-induced diabetic rats and in free radical scavenging activity. It was found to stimulate the release of insulin from pancreatic β cells and brought out improvement in lipid profile as well as in the architecture of pancreas and liver. HBAC thus opens the possibility of being developed as a combinatorial or monotherapy medicine for diabetes (Ahmed *et al* 2015). A recent study in China has revealed that the petioles and pedicels of *E. ferox* possess significant antioxidant activities (Wu *et al* 2014). Some water plants have been shown for their cosmeceutical ingredients that could help cure cutaneous aging (Lai et *al* 2015). For example, the seed extract of *E. ferox* appears as an effective cosmeceutical agent (Choo *et al* 2009). A recent study in Korea demonstrates the ethyl acetate fraction of *E. ferox* seed extract as a significant contributor to the inhibition of melanogeneis by altering liposomal degradation of tyrosinase (Baek *et al* 2015). This shows Makhana seed as a new cosmetic skin fairness agent. The hard seed coat of *E. ferox*, that is generally discarded, also comprises medicinal properties. A novel mucoadhesive biopolymer from *E. ferox* seed coat has been isolated and characterized which could be effective in drug delivery (Uma Shankar et *al* 2011).

A recent study in China has confirmed the hypoglycemic effect of triterpenoid-rich extract from *E. ferox* seed-shell on the normal and streptozotocin-induced diabetic mice (Yuan *et al* 2014). The hypoglycemic action was shown to occur through the regulation of glucose metabolism. In response to the treatment, the body weight of the experimental animals was shown to return to normal levels. Morphology of the islets of the pancreatic cells also recovered, as expected. There was a reduction in the expression of the negative regulation protein PTP1B gene with a concomitant increase in the insulin receptor IRS-1 protein expression. Anti-oxidant properties in *E. ferox* seed shell make it effective in preserving the pork sausage (Zhang *et al* 2015). Phenolic extracts from its seed coat have been found to have a significant anti-fatigue effect (Wu *et al* 2013). The starchy kernel of *E. ferox* has emerged as a tool in bone tissue engineering. A biomimetic, osteoinductive, tricomposite scaffold incorporating the starch of *E. ferox* seed with nanohydroxyapatite and chitosan have been synthesized. An in vitro synthetic bone scaffold study exhibited excellent biomineralization capacity of the nanocomposite made by nano-hydroxyapatite- chitosan -*E. ferox* starch that showed a comparatively better antibacterial activity against certain bacteria (Shakir *et al* 2018). A recent investigation on transcriptomic sequencing and analysis during growth of *E. ferox* seed and its development has revealed that the two vital differential genes PAL and P450 remain associated with phenylpropanoid biosynthesis. This indicates a relationship between morphological changes and accumulation of medicinal components in the plant (Liu et al 2018). Food technologists are now preparing nutri-bars for its possible use by the pregnant and lactating women. These are also consumed by women of childbearing potential who are on way to pregnancy. Nutri-bars are rich in vitamins that enhance the nutritional status of such women. In one such experiment, it was reported that a nutri-bar containing semolina (92.5g), when fortified with Makhana (7.5g), coconut (6g), cashew, almond, gum arabica (5g each), fenugreek, ginger (0.5g each) and honey (42g) yielded a good consistent, nutritional value and better organoleptic properties.

The property of floatability of Makhana seed powder has been exploited as an effective drug delivery agent. Makhana powder obtained after crushing its pops keeps on floating over the water surface and mixes only after continued stirring. This property of *E. ferox* has been involved in developing a drug delivery system. Based on this concept, a recent work has led to the development of a non-effervescent floating matrix tablet with *E. ferox* seed .

Recent years have witnessed a growing scientific attention in finding out the biochemical ingredient in E. ferox. Including bioactive metabolites (Jana *et al* 2022, Kumar *et al* 2024 etc..). Duly recognized as superfood , several products like pasta and laddoos are being prepared from it, either as single or blended product. (Devi *et al* 2024, Ekal and Kumbamoorthy 2024 etc..). Earlier Mishra *et al* (2014) prepared gluten free biscuits made of Makhana. Nutraceutical properties of Makhana have been further highlighted by Kapoor *et al* (2022), Rathod *et al* (2023), Jyothi and Bukya (2024), Shukla *et al* (2024), Vanishree *et al* (2024) etc.. Parray et al (2010) reported the methanolic extract of seeds and leaves of *E . ferox* to have antibacterial action against several pathogenic strains. Their findings substantiate the conventional use of *E. ferox* against kidney problems. Time is now ripe for deriving optimum productivity from waterlogged areas that grow a variety of aquatic fruits including *E. ferox*. (Singh *et al* 2023).

REFERENCES

1. Ahmed, D.; Sharma, M.; Kumar, V.; Bajaj. H.K. and Verma, A. (2015). 2β – hydroxy betulinic acid 3β caprylate: An active principle from *Euryale ferox* Salisb. seeds with antidiabetic, antioxidant, pancreas and hepatoprotective potential in streptozotocin induced diabetic rats. J. Food Sci. Technol., 52, 5427-5441.

2. Ahmed, D.; Kumar, V.; Verma, A.; Shukla, G.S.; Sharma, M. (2015). Antidiabetic, anti-hyperlipidemic effect of extract of *Euryale ferox* Salisb. with enhanced histopathology of pancreas, liver, kidney in streptozotocin-induced diabetic rats. SpringerPlus, , 4, 315.

3. Ahmed, D.; Kumar, V.; Verma, A.; Shukla, G.S. and Sharma, M. (2018). Antidiabetic, anti-hyperlipidemic effect of extract of *Euryale ferox* 554 Current Pharmaceutical Biotechnology, Vol. 19, No. 7

4. Baek, S.H.; Nam, I.J.; Kwak, H.S.; Kim, K.C. and Lee, S.H. (2015). Cellular antimelanogenic effect of a *Euryale ferox* seed extract ethyl acetate fraction via the lysosomal degradation machinery. Int. J. Mol. Sci., 16, 9217-9235.

5. Bana, M. and Gupta, R.K. (2015). Formulation, nutritional and phytochemical analysis of ready to mix infant food using Gorgon Nut, Samak Rice and Banana powder. J. Pharmacognosy Phytochem. 4, 76- 81

6. Bhowmik, S.; Dutta, B.K. and Saha, A.K. (2012). Determination of mineral content and heavy metal content of some traditionally important aquatic plants of Tripura, India using atomic absorption spectroscopy. J. Agri. Technol. 8, 1467-1476

7. Choo, S.J.; Kim, Y.H.; Ryoo I.J.; Xu, G.H. and Yoo, I.D. (2009). Application of a cosmeceutical ingredient of *Euryale ferox* seed extract. J. Soc. Cosmet. Sci. Korea, 35, 309-315

8. Das, S.; Der, P.; Raychaudhuri, U.; Maulik, N. and Das, D.K. (2006). The effect of *Euryale ferox* (Makhana), an herb of aquatic origin, on myocardial ischemic reperfusion injury. Mol. Cell. Biochem. 289, 55-63.

9. Jyothi, E. and Bukya, A. . 2024. Preparation and quality characterization of Makhana Sesame Laddu .

10. International Journal of Agriculture and Food Science (2024) 6(2): 11-15. DOI: https://doi.org/10.33545/2664844X.2024.v6.i2a.195

11. Ekal, V. and Silambuselvi K., (2024). Formulation and Proximate Analysis of Makhana-Enriched Laddoos International Journal of Nutrition, Pharmacology and Neurological Diseases 14(2):p 205-212, April-June 2024. | DOI: 10.4103/ijnpnd.ijnpnd_92_23

12. Hsu, H. and Cho, C. (1951). The nutritive value of Chinese formula "Su-shin". J. Taiwan Pharm. Assoc., 31, 25-28

13. Jan, K.; Jairajpuri, D.S. and Jan, S. (2012). Preparation of nutri-bar for lactating women. IOSR J. Env. Sci. Toxicol. Food Technol. (IOSRJESTFT). 1, 10-14.

14. Jana, B.R. , Bhatt B.P. and Srivastav, A. (2022). Biochemical components of Makhana (*Euryale ferox* Salisb) and its relevance to antiobesity effect in human. Asian Journal of Dairy and Food Research .Doi 10.18805/ajdfr.DR-1871

15. Jain, A.; Singh, H.B. and Kanjilal, P.B. (2010). Economics of foxnut (*Euryale ferox* Salisb.) cultivation: A case study of from Manipur in northeastern India. Indian J. Nat. Prod. Resour. 1, 63-67.

16. Jha, V. (2016). Makhana (*Euryale ferox* Salisb.): a significant nutraceutical. In: Indigenous Knowledge and Plant Utilization; Gupta, A, Ed.; Aavishkar Publications, Jaipur,; pp 253-267.

17. Jha, V.; Kargupta, A.N.; Dutta, R.N.; Jha, U.N.; Mishra, R.K. and Saraswati, K.C. (1991a). Utilization and conservation of *Euryale ferox* Salisbury in Mithila (North Bihar). Indian Aquat. Bot., 39, 295-314.

18. Jha, V.; Barat, G.K. and Jha, U.N. (1991b). Nutritional evaluation of *Euryale ferox* Salisb. (Makhana). J. Food Sci. Technol., 28, 326- 328

19. Jha, V.; Verma, A.B. and Jha, P. (2014). Aquatic macrophytes effective in control of diabetes: A review. Ann. Plant Sci., 3, 645-650.

20. Kapoor, S.,. Kaur A.,. Kaur, R., Kumar, V. and Choudhary, M.. (2022). *Euryale ferox*, a prominent superfood: nutritional, pharmaceutical, and its economical importance. DOI: 10.1111/jfbc.14435

21. Kim, Y.H.; Lee, M.J.; Lee, H.S.; Kim, J.G. and Park, W.H. (2011). Screening of antioxidative effect and suppressive effect of LDL oxidation of *E. ferox* Salisbury. Korean J. Oriental Physiol. Pathol., 25, 92-99.

22. Kumari, A. and Jha, V. (2017). Ethnic uses of Makhana (*Euryale ferox* Salisb.) in Mithila (north Bihar) and other parts of India. J. Trad. Folk Pract., 5, 60-74

23. Kumar,V., Kumar, A., Singh, M. K., , Dhyani, P., Mishra H. and Rai, D. C.. (2024). Bioactive metabolites identifcation of the foxnut and broken millet-based nutritional bar using HR-MS. Food Chemistry: Molecular Sciences, 9(2024) 100214

24. Kumari, A. (2013). Studies on Morphological and Cytological Characterization of Makhana Germplasm and its Ethnobotanical Aspects. Ph. D. Thesis, L. N. Mithila University, Darbhanga.

25. Kumar, S.; Singh, I.; Nandi, B. and Singh, C. (2014). Application and effect of addition of popped Makhana flour on the properties and qualities of bun. Int. J. Food and Nutri. Sci. 3, 116-120

26. Kumar, S.; Singh, I.; Kumar, N.; Singh, C. (2015). Application and effect of addition of popped Makhana flour on the properties and qualities of cookies. Int. J. Proc. Post-Harvest. Technol 6, 80-86.

27. Lai, W.F. and Lin, M.C. (2015). Treating cutaneous aging with patented technologies. J. Biosci. 40, 209-216.

28. Lee, M.R.; Kim, J.H.; Son, E.S. and Park, H.R. (2009). Protective effects of extract from *Euryale ferox* against glutamate-induced cytotoxicity in neuronal cells. Nat. Prod. Sci. 15, 162-166.

29. Liu, X.; He, Z.; Yin, Y.; Xu, X.; Wu, W. and Li, L. (2018). Transcriptome sequencing and analysis during seed growth and development in *Euryale ferox* Salisb. BMC Genomics, 19, 343.

30. Li, L.J.; Wu, Y.F. and Cao, B.S. (2007). Research progress of *E. ferox* seeds. China Vegetables, 81-88

31. Mishra A., Devi, M. and Jha, P. . (2014). Development of gluten-free biscuits utilizing fruits and starchy vegetable powders J Food Sci Technol Oct 9;52(7):4423–4431. doi: 10.1007/s13197-014-1521-5

32. Mridula D, R.K.Vishwakarma, Simran Arora and Manju Bala. (2024). Formulation and characterization of novel cereal glutenfree Pasta from semipopped Makhana, water chestnut and potato. Food and Bioprocess Technology 17:4262–4277 https://doi.org/10.1007/s11947-024-03349-5

33. Nath, B.K. and Chakraborty, A.K. (1985a). Studies on the amino acid composition of the seeds of *Euryale ferox* Salisb. J. Food Sci. Technol., 22, 293.

34. Nath, B.K. and Chakraborty, A.K. 1985b. Studies on the physico-chemical properties of the starch of *Euryale ferox*. Starch/Starke, 37,361-363

35. Negi, J.S.; Jugran, V. and Kasliwal, N. (2011). Development of noneffervescent floating matrix tablets based on *E. ferox* seeds. Asian J. Pharmaceut., 5, 93-100.

36. Parray, J.A.; Kamili, R.; Qadri, R.; Hamid, R. and De Silva, J.A.T. (2010). Evaluation of the antibacterial activity of *Euryale ferox* Salisb.- a threatened aquatic plant of Kashmir Himalaya. MAPSB, 2010, 4, 80-83.

37. Puri, A.; Sahai, R.; Kiran, L.; Singh, R.P. and Saxena, K.C. (2000). Immunostimulant activity of dry fruits and plant materials used in Indian traditional medicinal systems for mothers after childbirth and invalids. J. Ethnopharmacol., 71(1-2), 89-92.

38. Rathod, R.V., Neve, G., Jain, A. and Giri, P. (2023). A comprehensive review on health benefits and nutritional aspects of fox nut (Makhana). The Pharma Innovation Journal 2023; 12(6): 4432-4438

39. Row, L.C.; Ho, J.C. and Chen, C.M. (2007). Cerebrosides and tocopherol trimers from the seeds of *E. ferox*. J. Nat. Prod., 70, 1214- 1217.

40. Shamim, N. and Paul, V. (2017). Utilization of flours of foxnuts and water chestnuts for preparation of Pua. Int. J. Adv. Res. 5, 880-883

41. Shakir, M.; Mirza, S.; Jolly, R.; Rauf, A. and Owais, M. (2018). Synthesis, characterization and in vitro screening of a nano- hydroxyapatite/ chitosan/ *Euryale ferox* nanoensemble- an inimitable approach for bone tissue engineering. New J. Chem., 42, 363-371.

42. Sharma. K.; Patel, S., Devi, M. and Vishwakarma, R.K. (2024). Effect of packaging material and storage period on popping behaviour of roasted Makhana seeds. Journal of Agricultural Engineering (INDIA), Vo 61 (4) DOI: https://doi.org/10.52151/jae2024614.1870

43. Shukla, N., Sharma, J.and Sharma, P. K. (2024). A review on: *Euryale ferox* (fox nut). 2024 Journal of Pharmacognosy and Phytochemistry. 13(5): 50-52

44. Shen, B.; Qinan, W.U.; Rong, C.; Wei, Y.U.E. and Xiaoxiao, H.E. (2012). The modern research progress of *Euryale ferox*. Northwest Provincial J. (abstract) (02).

45. Singh, P.K. Distribution and uses of Makhana in Manipur. (2003). In: R.K. Mishra, V.; Jha and P.V. Dehadrai (eds.) MAKHANA, DIPA, ICAR, New Delhi, pp 8-13.

46. Singh, M.R. and Gupta, A. (2006). Cultivation and conservation practices of *E. ferox* Salisb. in Manipur. Indian J. Trad. Know. 5, 143-144.

47. Singh, H. K., Singh, P., Singh, R., Kumar, A. and Kumar, A. (2023). Performance of Makhana (*Euryale ferox* Salisb.) as a highly profitable venture in waterlogged areas prone to flooding. International Journal of Statistics and Applied Mathematics SP-8(4): 653-656 DOI: https://doi.org/10.22271/maths.2023.v8.i4Si.1133

48. Song, C.W.; Wang, S.M.; Zhou, L.L.; Hou, F.F.; Jwang, K.; Han, Q.B.; Li, N. and Cheng, Y.X. (2011). Isolation and identification of compounds responsible for antioxidant capacity of *Euryale ferox* seeds. J. Agric. Food Chem., 59, 1199-1204.

49. Song, J. and Wu, Q.N. (2009). Analysis of therapeutic effects of Qian Shi on proteinuria. China J. Mod. Drug Appl. 3, 133-134.

50. Uma Shankar, M.S. and Sateesh Madhev, N.V. (2011). Isolation and characterization of a novel smart mucoadhesive bioadhesive biopolymer from *Euryale ferox* seed coat. Pharmabit, 23-24 (1-2), 57-60.

51. Vanishree, M., Karthikeyan, G B. and Prakash, D. A. (2024) Fox nut (Makhana): the power of unknown seed. International Journal for Multidisciplinary Research (IJFMR) E-ISSN: 2582-2160

52. Wu, C.; Chen, R.; Wang, X S.; Shen, B.; Yue, W. and Wu, Q. (2013). Antioxidant and anti-fatigue activities of phenolic extract from the seed coat of *Euryale ferox* Salisb. and identification of three phenolic compounds by LC-ESI-MS/MS. Molecules, 18, 11003- 11021.

53. Wu, C.; Wang, X.; Wang, H.; Shen, B.; He, X.; Gu, W. and Wu. Q. (2014). Extraction optimization, isolation, preliminary structural characterization and antioxidant activities of the cell wall polysaccharides in the petioles and pedicels of Chinese herbal medicine Qian (*Euryale ferox* Salisb.). Int. J. Biol. Macromol. 64, 458-467.

54. Zhang, S.; Cheng, H. and Dong, J. (2011). Amino acid and mineral composition of the seeds of *Euryale ferox*. Chem. Nat. Compounds, 47, 490-491

55. Zhang, C. (2015). In vitro antioxidant properties of *Euryale ferox* seed shell extracts and their preservation effects on pork sausages. J. Food Process. Preservat. 39, 1172-1182

56. Zhang, R. and Cui, Z.M. (2009). The advance of *E. ferox* Salisb. fundamental and application research. Chin. Sci. Technol. Inf., 26, 130-132.

57. Zhao, H.; Zhao, S.; Sun, C. and Guillaume, D. (1989). Glucosylsterols in extract of *E. ferox* identified by high-resolution NMR and mass spectrometry. J. Lipid Res., 30, 1633-1637

58. Zhao, H. and Zhao, S. (2007). New cerebrosides from *E. ferox*. J. Nat. Prod., 57, 138-141

59. Zhiwei, W.Y.; Xiaoyan, Y.; Quansheng, G. and Meixin, J. (2011). Morphological characteristics of starch grains of root and tuber plants in South China. Quartern. Sci., 31, 736-745.

15

Recent Investigations for Development of Makhana Crop and its Products

Rachana Shalini[1] and Vidyanath Jha[2]

[1]*Dept. of Botany, J.N. College, Madhubani, Bihar*
[2] *L.N.Mithila University, Darbhanga, Bihar*

INTRODUCTION

The foremost amongst the aquatic plants growing in the water bodies of north Bihar is *Euryale ferox* Salisb. Basically it is reported to be a perennial, rooted, floating plant belonging to the family *Nymphaeaceae*. However, in north Bihar it is at the most an annual plant as the plants have to be uprooted to allow its natural disintegration that ultimately facilitates its harvest. Its seeds are collected virtually by sweeping the floor by holding deep breaths inside the water. The fishing community in this area has developed the acumen of the intricate methods of cultivation, harvest and processing through generations. As such, they have a monopolistic expertise over the Makhana crop.

Recent decades have witnessed an expansion of Makhana cultivation in the water-logged areas of Assam, West Bengal, Odisha, and Manipur etc. However, it is only the fishermen from north Bihar who are hired for performing the harvest. It is extremely difficult for a novice to take up the harvest of its fruits and seeds as the plants of *E. ferox* are loaded with sharp prickles and it is extremely difficult to even touch them.

Makhana crop-cycle, harvest practices and recent interventions *E. ferox* is an eight month crop that is generally grown in shallow water bodies having non-calcareous soil. Eight to ten districts of north Bihar in Darbhanga, Kosi and Purnea divisions could be considered as major areas of its cultivation. Darghanga, Madhubani, Katihar and Purnea are the major centres of Makhana processing.

It is a rooted floating plant having giant sized leaves. It is called as Asiatic cousin of *Victoria*. Its entire body parts except root are covered with prickly spines and as such it is very difficult for a person to enter into a Makhana pond unharmed. It is on account of this fact that the plants have to be uprooted and destroyed to enable the harvest of its seeds from the pond bottom.

The plant undergoes asynchronous flowering. The solitary flowers are of two types (a) cleistogamous that does not open and remains submerged throughout. The other type of flower is of (b) chasmogamous type that comes into the air, remains open for two to three days and then retrieves back into submergence under the leaves. Both the types of flowers undergo seed setting. However, there is predominance of seed setting in cleistogemons flower. Flowers start appearing by late May. The bursting of fruits starts during June onwards. The seeds shed remain floating over water surface for a few days till the aerenchymatous aril shrinks and flatters. Seeds remain at the pond bottom and have to be harvested from there. Wild forms of *E. ferox* do have a perennial nature but its domestication as a crop has registered it to remain an annual crop. Makhana system also provides an opportunity for integrated aquaculture. Makhana ponds are ideal habitats for the growth of air breathing fishes like Singhi (*Heteropneustes fossilis*), Mangur (*Clarias batrachus*), Kabai (*Anabas testudineus*) etc. These detritivorous fishes thrive in the heavy organic debris that gets deposited at the pond bottom.

A fresh pond in which Makhana is to be grown needs to be sown with freshly collected raw seeds from another site. Seeding is not required in a pond in which Makhana has been already growing. This is because of the fact that how- so-ever efficiently harvests have been made some seeds are found to remain at the bottom. Such seeds sprout during January- February to constitute the next crop.

There is also a practice of making transplantation of the crop during the second half of April. This helps the crop escape the competition with the other aquatic weeds that grow along with Makhana. Normally these weeds are stacked as heaps in the pond itself. These heaps disintegrate during the course of time and act as manures to the main crop. This is an example of nutrient recycling that benefits the crop.

Maintenance of water level in Makhana ponds is a major limiting factor. For about 5-7 years this region has evaded the onset of normal monsoon rains. As such, the Makhana ponds are witnessing a drought like situation and it is proving difficult for the marginal farmers to keep a minimum water depth of almost the knee-dip water that is essential for the crop to thrive upon.

Harvest starts during August-September and continues up to the onset of winter during November- December. The harvested seeds are processed under an intricate method phase wise as follows:-

1. Sun drying and storage
2. Grading of raw seeds with sieves
3. Preheating

4. Tempering
5. Roasting
6. Popping
7. Separation
8. Grading of pops with sieves
9. Removal of left over seed- coats and Polishing
10. Packing of pops for marketing.

DISCUSSION

Scientific studies on post harvest technology of gorgon nut have been made by Jha and Prasad (2003). The manual method of post harvest processing is usually undertaken by the women of fishing community and is quite cumbersome. A processing machine has been developed by the Ludhiana based CIPHET (Central Institute of Post-harvest Engineering Technology). It was earlier designed at the Agriculture and Food Engineering Department of the I.I.T. Kharagpur. Its recent version is an improved one and yields more than 90% of the grade one Makhana pop. Recently the local traders in Makhana business have installed a polishing machine that helps removing the left over seed-coats and adding luster to the pop. This increases their sale value. Pops are packaged into different sizes with the help of machine and are marketed. No sand is applied while roasting / popping Makhana seeds manually; while the same is essential in case of Khubahi. Sand protects Khubahi seeds from coming in direct contact with the heat of earthen pot. Raw Makhana seeds have hard seed-coats and also a higher kernel mass. As such it does not face the danger of disfigurement and consequent damage by the high heat.

An abbreviated account of Makhana cycle, harvest practices and recent interventions have been provided in tabular form in Table –1.

Table-1: Makhana Cycle, Harvest Practices and Recent Interventions

Components	Details
Crop –cycle	January to July-August (Sown crop) April to August-September (Transplanted crop)
Harvest	Generally performed during August- October with bamboo appliances. Makhana harvesters from Darbhanga and Madhubani districts migrate to Maldah area of West Bengal, Odisha and Assam.
Popping	September –December. A machine developed by CIPHET, Ludhiana works perfectly in production of good quality pops.

Components	Details
First ever variety released	'Swarna Vaidehi' released by Research Centre for Makhana, Darbhanga under ICAR-RCER set up
Integrated aquaculture	Makhana habitat suitable to the growth of air breathing fishes. Work done recently on integration with carp fishes by leaving open spaces.

Contrivances Used in Makhana Harvest

Makhana harvesters uses some bamboo gears that are produced indigenously.

- Auka is utilized for stocking the harvested raw seeds from the pond bed to the shore from August to October. It is a cylindrical large bamboo bucket with one end closed having 1.5m-2m in height and 1m-1.5m in diameter.
- Gaanja is used along with Auka. It is an inverted bell or comma shaped structure having 50 cm with 35 cm diameter at the rim portion.
- Kaara are the bamboo poles fixed in the water bodies for demarcating the areas to be swept by individual groups of 7-8 harvesters. Collected seeds at the bottom are heaped around these Kaaras.
- Khainjhi is also a cylindrical smaller bamboo bucket like Auka utilized for storing raw and popped seeds.

Apart from these bamboo gears, some other indigenous appliances are also utilized by Makhana harvesters.

Pitcher floats or gharra is used for movement in the interiors of pond in harvesting Makhana and Singhara. It is made up of two earthen pitchers tied together in inverted position with the help of two bamboo sticks. Ropes normally interwoven from paddy-straw or coir is used to control aphids. It is held on the two ends and randomly rolled over the tender leaves. It is also used for clearing the unwanted weeds before transplantation of the crop. Vines of *Ichnocarpus frutescens* (Dudhlatti) and *Tiliacora acuminata* (Kariyauti) are utilized as cordage in bamboo appliances (Jha et al. 1998). Small boats made from the wood of *Dalbergia sissoo, Syzygium cumini* and *Mangifera indica* are used for seeding, transplantation of seedlings, spray of insecticides and transfer of harvested seeds off shore. Mats are used for sun drying the raw seeds before roasting. A mat has a size of 2-2.5 to 1-1.5m and is generally made up of *Desmostachya bipinnata* (Kush), *Phoenix dactylifera, Typha* sp. and *Cyperus* sp. Sieves of different mesh sizes ranging from 0.5cm to 1.25cm are used for gradation of raw seeds and pops. These are made up of iron sheets with wooden frame. Aphara and Thapi are made up of the heart wood of *Dalbergia sissoo, Shorea robusta* and *Acacia lenticularis.* Aphara is the wooden platform on which roasted seeds (Guris) are hit with the flat wooden hammer called Thapi. Thick earthen pans are used for roasting the seeds. Bamboo baskets, earthen pitchers, aluminium utensils and jute or plastic bags are used to store the raw seeds and pops of Makhana.

Interventions to reduce drudgery in Makhana harvest and processing:

Makhana harvest is performed by male members of the fishing community by holding deep-breaths inside water. Since Makhana harvest and processing mechanism are quite cumbersome, efforts are being made to make the exercise less painstaking. As such mechanical interventions have been made in recent times. These are aimed towards bringing down the discomforts encountered during the process. Very recently Khadatkar et al in 2015 have suggested an improved system of its harvest that consists of a floating platform providing support to a 10 liter cylinder having compressed air, 10m hose pipe with regulator and a mini diving kit having suit with cap, mask and content guage. This system brings down the overall discomfort rate from 8.3 in the traditional system to 4.2 in the case of the improved one. The body part discomfort score also comes down from 78.8 in the traditional system to 48.2 in the improved system. Average output of seed procurement in the traditional system is 3.8 kg/hr which increases to 11.3 kg/hr in the improved system. Thus the improved system involves less drudgery in comparison to the traditional one with significantly higher work output.

Drudgeries are also experienced during pop making, which is done at a very high surface temperature i.e. 290-340^0 C. Facing such a high heat is highly dangerous. As such, CIPHET (Central Institute of Post Harvest Engineering and Technology), Ludhiana has recently made an intervention in the form of development of a machine for popping and decortication of Makhana seeds. It has done away with the tedious conventional method of manually roasting and hitting the hot roasted nuts by mallets. The new system facilitates continuous roasting and popping operation. The new machine consists of a screw conveyer type roaster and a centrifugal type popping unit that can also decorticate the conditioned seeds. This was not in practice earlier. The developed machine is capable of both popping and decorticating the gorgon nut, if properly conditioned. The capacity of the developed system is about 40 kg of raw seeds which can be easily scaled up or down. The machine performs excellently and gives 90 % efficiency. It was tested and demonstrated among Makhana farmers and traders at Research Centre for Makhana on Dec. 2013. Quality of popped Makhana is much better than the one obtained through traditional method.

Intervention has also been made for development of Makhana grader that consists of main frame, sieve unit and operator unit as three major components. The machine grades into four categories at crank speed of 45 rpm and sieve angle of 15 degrees at a capacity of 700-750 kg/hour. Two persons are required for smooth operation of the grader, one for pedaling and another for filling Makhana seeds from the sieve mouth.

In the traditional method roughly 54% pops are of grade one type called *Rasgulla*, followed by 35% and 11% respectively of the medium quality *Murra* and the low grade *Thurri* types. This handicap has now been overcome with the development of a machine which has achieved perfection in the process of pop making. The machine has been developed by CIPHET, Ludhiana. The new machine yields over 90% of the grade one Makhana pops. However the problem remains with the broken seed coat still remaining attached with the pops. The left-over hard seed coats create problem during the intake of Makhana products either in the form of snacks or desserts. Earlier black seed coats were removed manually. This difficulty has now been done away with the introduction of Makhana polishing machine by the local traders some five years back. This machine removes these leftover seed coats and adds luster to the pops. This increases the sale value and quality of the finished produce.

Nutritional Properties of Makhana:

Makhana is in possession of significant nutritional and medicinal properties that have been summarized in Table – 2. It provides the details of the nutritional characteristics of *E. ferox* seeds in both raw and popped forms. Makhana does have lower protein content (10-12%) as compared with most cereals. But it is nutritionally superior to several plant and animal-based diets. The essential amino acid indices (EAAI) in the raw and popped Makhana were found to be 93% and 89% respectively, which are higher than the values of plant-based diets like rice, wheat, Bengal gram, soya beans and amaranth and animal-based diets like human and cow's milk, fish and mutton.

Values of the chemical score also were higher as compared with most of the plant based diets. Makhana has a higher amount of alanine, tyrosine and arginine. Popping brings about a decline in amounts of tyrosine, histidine, aspartic acid, proline and methionine. In Makhana (A + L)/P ratio in food explains the rate of protein utilization for growth.

In Tripura, India, wild population of *E. ferox* was found to have a higher value of this ratio. A low biological value (BV) could be accounted for to the higher ratios of Leucine to Isoleucine in its seeds. It is because of its lower BV that Makhana is recommended as a complementary food item along with the staple ones.

Table-2: Nutritional Properties of Makhana

Parameters	Content
Seed Kernel	
Protein	9.7%
Carbohydrate	79%
Fat	0.1%
Iron	1.4mg/100g
Calorific Value	
Raw seed kernel	362kcal/100g
Popped form	328kcal/100g
Essential Amino Acid Index	

Parameters	Content
Raw seed kernel	93.63%
Popped form	89.95%
Biological Value	
Popped form	55
Chemical Score (% Egg)	
Raw seed kernel	70%
Popped form	56.57%
Comparative values of (a). Leucine/Isoleucine	
Raw seed kernel	1.84
Popped form	1.99
(b). (Arginine+Lysine)/Proline	
Raw seed kernel	6.30
Popped form	4.74

Perisperm in the seed of *E. ferox* is its edible part. Raw seed is converted into popped form through an arduous method of post-harvest processing. It is an integral component of a number of ethnic practices in Mithila and other parts of India. The traditional Ayurvedic system of Indian medicine holds Makhana as beneficial in 'Vata' and 'Pitta' disorders. It acts as an expectorant, emetic and as a cardiac stimulant. Both seeds and flowers have aphrodisiac properties. Ayurveda mentions it as spermatogenic. Due to this property of Makhana in Mithila region of north Bihar it is customary for a new bride's parents to send Makhana to the grooms on the occasion of Kojagara festival (a marital ritual). Makhana is a significant nutraceutical and is a good source of micronutrients. A ready-to-mix infant food using a gorgon nut, *samak* rice and banana powder has been formulated.

A recent study in China has revealed that the petioles and pedicels of *E. ferox* possess significant antioxidant activities. *Euryale ferox* is used in the Korean traditional medicine for the control of dysentery, diarrhoea, leucorrhoea, incontinence and paralysis of joints. Makhana seeds are also understood to possess immunoral properties. A number of other global studies have mentioned several medicinal aspects of this plant. These studies suggest that this aquaphyte has good prospects of preventing cardiocerebral diseases, therefore, a good source material for developing new drugs and health products including flavonoids, cyclic peptide, sterol, brain lipid *etc.*

A new compound 2β –hydroxy betulinic acid 3β caprylate (HBAC) isolated from *E. ferox* seeds has been found to be effective in glycemic control in streptozotocin-induced diabetic rats and in free radical scavenging activity. It was found to stimulate the release of insulin from pancreatic β cells and brought out improvement in lipid profile as well as in the architecture of pancreas and liver. HBAC thus opens the possibility to be developed as a combinatorial or monotherapy medicine for diabetes. The seed extract of *E. ferox* appears as an effective cosmeceutical agent. A recent study in Korea demonstrates the ethyl acetate fraction of *E. ferox* seed extract as a significant contributor to the inhibition of melanogeneis by altering liposomal degradation of tyrosinase. This shows Makhana seed as a new cosmetic skin fairness agent.

A recent investigation on transcriptomic sequencing and analysis during growth of *E. ferox* seed and its development has revealed that the two vital differential genes PAL and P450 remain associated with phenylpropanoid biosynthesis. This indicates a relationship between morphological changes and accumulation of medicinal components in the plant.

The hard seed coat of *E. ferox,* that is generally discarded, also comprises medicinal properties. A novel mucoadhesive biopolymer from *E. ferox* seed coat has been isolated and characterized which could be effective in drug delivery. Phenolic extracts from its seed coat have been found to have a significant anti-fatigue effect.

Food technologists are now preparing nutri-bars for its possible use by the pregnant and lactating women. These are also consumed by women of childbearing potential who are on way to pregnancy. Nutri-bars are rich in vitamins that enhance the nutritional status of such women. In one such experiment, it was reported that a nutri-bar containing semolina (92.5g), when fortified with Makhana (7.5g), coconut (6g), cashew, almond, gum arabica (5g each), fenugreek, ginger (0.5g each) and honey (42g) yielded a good consistent, nutritional value and better organoleptic properties.

Dietary Formulations Using E. ferox as a Component

Makhana could be held as a significant nutraceutical. It is being used in development of drugs, often in combination with other plant products. A number of formulations, that incorporate *E. ferox* as one of the ingredient exist. These include (a). Nutri-bar developed in India, (Jan, K.et. al., 2012) (b). Yishen capsule in China (Fang, J., 2003; Fang, J. et. al. 2003 4(7); Fang, J. 2005; 2013) and (c). Herbal suppository, developed in Korea (Liang, K.; Liang, L. 2002).

A nutri-bar developed by Jan et al. in 2012 contains Semolina (92.5g), which is fortified with Makhana (7.5g), coconut (6g), cashew, almond and gum Arabica (5g each), fenugreek and ginger (0.5g each) and honey (42g). This combination yields a good consistency. It is high in nutritional value and has better organoleptic properties. Nutri-bar is rich in vitamins and enhances the nutrition status of pregnant and lactating women. These are also consumed by women of child bearing potential who are attempting to become pregnant. Yishen Capsule is a Chinese formulation that incorporates *E. ferox* as one of the ingredients. It has been found effective in control of diabetic nephropathy. It has been found to be effective in raising the level of podocalyxin protein that helps the glomerulus in kidney maintain its blood purification capacity. The herbal suppository (US 20020031559 AI, 14 March 2002, Kin Liaug and Liyin Liang) has a potential anti-allergenic property and used for treating undesired symptoms of allergic rhinitis, sinusitis, nasal congestion, nasal dripping, nasal polyps, infections, fever, cough, spasms, dizziness, convulsions etc..

Floatability in Makhana Seeds, Pops and Powder:

E. ferox has a unique property of floatability in its raw seeds, pops as well as powder obtained after crushing. The seeds released after the fruit dehiscence, keep on floating for a few days on the water surface. It is because of the presence of aril over the seeds that are filled with air. The seeds sink to the bottom after the shrinkage of aril. The aril is otherwise also a good source of nutrients.

The property of floatability of Makhana seed powder has been exploited as an effective drug delivery agent. Makhana powder obtained after crushing its pops keeps on floating over the water surface and mixes only after continued stirring. This property of *E. ferox* has been involved in developing a drug delivery system. Based on this concept, a recent work has led to the development of a non-effervescent floating matrix tablet with *E. ferox* seed.

E. ferox and Intellectual Property Rights:

Last two to three decades have witnessed a spurt in the investigations on this plant (Table 4). Increasing stress by social and academic groups led to the establishment of NRC for Makhana under ICAR-RCER set up. This centre has established the first-ever variety of Makhana *e.g.* 'Swarna Vaidehi'. The Variety Release Committee of the institute recently approved the release of Makhana variety for Bihar, Assam, Chhattisgarh and Odisha. The variety developed through pure line selection has the production potential of 2.8-3.0 t/ha. It has novelty in seed size and production. Seeds are deep black in colour, bold in size (62-106 g/fruit) and spherical in shape, 65-110 seeds/fruit, 12-16 fruits/plant. Popped seeds are big in size and sparkly white in color. Average yield is 2.8-3 t/ha in farmer's field which is almost two-fold higher than the productivity of traditional cultivars. Fruits (fully developed) are big in size and spheroidal in shape having whitish brown outer skin. The institute has also developed the technology for Makhana cultivation in the cropping system, wherein it is grown at a shallow water depth of just about 30cm. After Makhana, wheat, berseem or other rabi crops like lentil, gram or vegetables could be cultivated in the same field. 'Swarna Vaidehi' could be grown successfully in the stagnant water bodies and also in the field method of Makhana cultivation. Patents on Makhana in India have been applied for (i) a process for making storage edible food materials from kernels of gorgon nuts as well as on (ii) a process of manufacturing mix for ready to consume Makhana Kheer (Pudding). In 2002, a patent on a process of making storage edible food materials from kernels of gorgon nuts (IN patent 187500, 4 May 2002) was obtained.

The other Indian patent refers to the process for manufacture of well-standardized ready-to-consume Makhana Kheer mix (RCMK) with prolonged shelf life under suitable storage conditions. The method in the invention could be used to prepare Makhana kheer almost instantly by mixing one part of RCMK with desired parts of warm/hot/cold water. The new method ensures consistency of the Makhana Kheer and allows its wide commercialization due to long shelf life. Ground popped seed, whole milk powder or dairy whitener and sugar are the major constituents of the mix, with optional ingredients to enhance taste, flavour, colour and a binder to make a homogeneous mass during thorough blending or mixing. A Chinese patent entitled:- "*Euryale ferox* non-fried instant noodle" refers to an invention of a noodle comprising 30 to 50% of *Euryale ferox* seed powder, 62-34% of flour, 5-10% of starch, 1-2% salt and 2-4% of edible vegetable oil. *E. ferox* powder is added to the main raw materials of instant noodles. The vitamins and minerals in the noodles are claimed to guarantee nutritional components needed by a human body. It has the characteristics of mild tonification, increasing resistance to dryness and facilitating the loss of greasiness. It is supposed to invigorate the function of spleen and enhance the excretion of urine xylose and increase the concentration of carotene in serum. *E. ferox* noodle is claimed to make up the defect of existing instant noodles in terms of deficiency of nutritional components.

An American Patent (US 200 20031559 A1) (Publication date 14 March 2002 with 22 claims) (Inventers Kin Liaug, Liyin Liang) is on the anti-allergenic property in *E. ferox* (in association with a number of other plants) for treating undesired symptoms for allergic rhinitis, sinusitis, nasal congestion, nasal dripping, nasal polyps, infections, fever, coughs, spasms, dizziness, convulsions *etc.* A Chinese formulation called Yishen capsule, containing *E. ferox* as one of the ingredients, has been found effective in the control of diabetic nephropathy. *E. ferox* is also a component of a herbal composition to treat airway inflammation like allergic conjunctivitis (US 2013095171-A1 & TW 201322991). A paper published in the CSIR journal on IPR issues has analyzed the impacts of a patent on *Euryale ferox* on biodiversity at the micro level. The study has revealed that mass culture of Makhana in the two water bodies of Harishchandrapur- II block of Maldah district of West Bengal has created an interlinked ecological, social and economic imbalance. They have suggested a sustainable culture of both fish and Makhana in a manner that does not damage the local biodiversity, and the equilibrium of economy and ecology is sustained.

Table-4: Patents granted on *E. ferox*.

S. No.	Name of the Patent	Patent No.	Country
1	Kheer-mix	Application No.-579/ DEL/2001A	India
2	Storage-edible food material from kernels	IN Patent 187500. 4 May 2002	India
3	A mechanical system for popping and decortication of Makhana seeds	Application No. 674/ DeL/2013A, date of pub. 12/09/2014	India
4	Antiallergenic properties of *E. ferox*	US 20020031559A1 Pub. Date 14 March, 2002	America
5	Herbal composition to treat airway inflammation like allergic conjunctivitis	US 2013095171-A1 TW201322991	America
6	Calorigenic protein expressing promoter	JP 2003113100	Japan
7	Non-fried instant noodle	China Patent no. 20101053539620	China
8	Dehulling cutter for Gordon Euryale seed	Publication No. CN101912144A/2010	China

Chaitanya and Soam, (2011) made a Geographical Indication Portfolio Analysis for enhancing the income of Makhana growers in the state of Bihar. They are of the opinion that GI status in case of Makhana be granted to Bihar on the basis of extensive cultivation practices and fulfillment of country demand by 80-85 % by a single state. This would not only strengthen the livelihood and income generation to growers but will fortify the economy of the state as well.

Shankar and Madhav (2011) were able to isolate and characterize a novel mucoadhesive biopolymer from these seed-coats that could be effective in drug delivery mechanism. These seed-coats have a therapeutic effect on diabetes by insulin resistance. A recent study by Yuan et al. (2014) in China has recently confirmed the hypoglycaemic effect of triterpenoid rich extract from the seed-coat on normal and streptozotocin induced diabetic mice. Very recently, Kumar and Jena in 2015 have prepared high surface area microporous activated carbon from the seed-shell of *E.ferox.*

Considering the nutritional, medicinal as well the economic value of makhana, exploitation of quality specific traits is imperative, particularly the genotype specificity. Consequently, the identification of various value added traits may provide a platform for crop improvements in makhana. Kumar N. et.al. (2017) were recently investigated the presence of genetic diversity in germplasm of *E. ferox* with the help of SSR marker analysis, which could help in its genetic improvement. For this study germplasm were collected from different geographical states of

India and further assessed by SSR markers. This study could help in conservation of wild germplasm of *E. ferox* as well as in identification of different parental germplasm to widen the genomic base of this underutilized aquaphyte.

Makhana crop has a bright prospect of transforming the region of north Bihar as an economically sound hub for employment generation and consequent check on population exodus. New young entrepreneurs and researchers may take up the challenges and probabilities of launching all aspects of research and development based on scientific culture, harvest, production, relevant food processing, packaging and marketing of makhana. A lot of research work is required to minimize the drudgery during harvest and processing, ready-to-consume food products and different new nutraceutical and therapeutic uses of makhana. Some recent investigations demonstrate that Makhana is richer in the diversity of nutrients. Keeping in view its immense values as mentioned above, it is necessary to speed up research work on raising climate-resilient *Euryale ferox*, particularly through tissue culture and biotechnology

REFERENCES

1. Ahmed, D.; Sharma, M.; Kumar, V.; Bajaj. H.K.; Verma, A. 2β – hydroxy betulinic acid 3β caprylate: An active principle from *Euryale ferox* Salisb. seeds with antidiabetic, antioxidant, pancreas and hepatoprotective potential in streptozotocin induced diabetic rats. *J. Food Sci. Technol.*, 2015, *52*, 5427-5441.

2. Bhowmik, S., Dutta, B.K. and Saha, A.K. (2012). "Determination of mineral contents and heavy metal content of some traditionally important aquatic plants of Tripura, India using atomic absorption spectroscopy". Journal of Agricultural Technology. 8(4): 1467-1476.

3. Bana, M.; Gupta, R.K. Formulation, nutritional and phytochemical analysis of ready to mix infant food using Gorgon Nut, Samak Rice and Banana powder. *J. Pharmacognosy Phytochem.*, 2015, *4*, 76- 81.

4. Baek, S.H.; Nam, I.J.; Kwak, H.S.; Kim, K.C.; Lee, S.H. Cellular antimelanogenic effect of a *Euryale ferox* seed extract ethyl acetate fraction *via* the lysosomal degradation machinery. *Int. J. Mol. Sci.*, 2015, *16*, 9217-9235.

5. Choo, S.J.; Kim, Y.H.; Ryoo I.J.; Xu, G.H.; Yoo, I.D. Application of a cosmeceutical ingredient of *Euryale ferox* seed extract. *J. Soc. Cosmet. Sci. Korea*, 2009, *35*, 309-315.

6. Chaitanya, A. and Soam, S.k. Geographical Indication Portfolio Analysis for Enhancing Income of Makhana Growers in Bihar. Project Report, ICAR-National Academy of Agricultural Research Management, Hyderabad. 2011, 145 Pages.

7. Choudhuri, U.R.; Bhagat, P. A process for making storage-edible food materials from kernels of gorgon nuts. IN Patent, 187500, 4, 2002.

8. Chaudhuri, S.K.; Dutta, C. Impacts of a patent on *Euryale ferox* on biodiversity at micro level: A case study. *J. Intellect. Property Rights*, 2006, *11*, 430-435.

9. Fang, J. Clinical investigation of Yishen capsule in 78 elderly patients with diabetic nephropathy. *Chin J. Integ Medi Cardio/Cerebrovasc Dis.* 2003, 1(8): 454.

10. Fang, J., Wu, M. and Li, J. Observation Yishen Capsule on 35 cases of the membraneous proliferative glumoro nephritis. *Chin. J. Integ Trad. Western Nephrol.* 2003 4(7): 413.

11. Fang, J., Deng, A. and Liu, J. A clinical study of Yishen Capsule in treating early diabetic nephropathy. *Chin. J. Integ. Trad. Western Nephrol.* 2005, 6(8): 457-468.

12. Fang, J., Wei, H., Sun, Y., Zhang, X., Liu, W., Chang, R., Wang, Gong, Y. Regulation of podocalyxin expression in the kidney of streptozotocin –induced diabetic rats with Chinese herbs (Yishen Capsule). *BMC Complementary and Alternative Medicine.* 2013, 13: 76.

13. Guo, R.A.; Tang, X.H.; Sun, D.J.; Li, Z.F.; Liu, C.Y. Case study of thirty seven patients with proteinuria inhibitor by Qianshi. *J. Shangdong, Coll. Tradit. Chin. Med.*, 1993, 17, 32-33.

14. Jha, S.N.; Prasad, S. Post-harvest technology of gorgon nut. In: R.K. Mishra, V. Jha and P.V. Dehadrai (eds.) MAKHANA. Directorate of Information and Publications of Agriculture. *Indian Council Agricult. Res.*, New Delhi, 2003, 194-214.

15. Jha, V.; Verma, A. M. and Jha, A. K. (2003). "Indigenous contrivances utilized in *Makhana* cultivation in north and north-eastern India". In: R. K. Mishra, Vidyanath Jha, P. V. Dehadrai (eds.) MAKHANA; Indian Council of Agricultural Research, New Delhi.).

16. Jha, V.; Barat, G.K.; Jha, U.N. Nutritional evaluation of *Euryale ferox* Salisb. (Makhana). *J. Food Sci. Technol.*, 1991b, *28*, 326-328.

17. Jha, V. Makhana (*Euryale ferox* Salisb.): a significant nutraceutical. In: *Indigenous Knowledge and Plant Utilization;* Gupta, A, Ed.; Aavishkar Publications, Jaipur, 2016; pp 253-267.

18. Jan, K.; Jairajpuri, D.S.; Jan, S. Preparation of nutri-bar for lactating women. *IOSR J. Env. Sci. Toxicol. Food Technol.* (IOSRJESTFT), 2012, *1*, 10-14.

19. Jha, S.N. Mechanised system for popping and decortications of Makhana seeds (gorgon nut/*Euryale ferox*) Patent application publication, application number 674/DEL/2013A, Date of filing 07/03/2013, publication date 12/09/2014. The Patent Office Journal 12/09/2014.

20. Khadatkar, A.; Gite, L. P. and Gupta, V. K. (2015). "Interventions to reduce drudgery of workers in the traditional method of harvesting Makhana (*Euryale ferox* Salisb.) seeds from ponds". Current Science. 109(7): 1332-1337.

21. Kumari, A.; Jha, V. Ethnic uses of Makhana (*Euryale ferox* Salisb.) in Mithila (north Bihar) and other parts of India. *J. Trad. Folk Pract.*, 2017, *5*, 60-74.

22. Kumar, N.; Shikha, D.; Kumari, S.; Chaudhary, B.K.; Kumar, L.; Singh, I.S. SSR-Based DNA fingerprinting and diversity assessment among germplasm of *Euryale ferox:* an aquatic underutilized and neglected food crop. *Applied Biochemistry and Biotechnology,* 2017, 185, 1-8.

23. Kumar, A.; Singh.; I.S.; Thakur, A.K.; Choudhary, A.K.; Jha, V.; Singh, S.P.; Prasad, S.S.; Yadav, P.; Kumar, R. Bioaccumulation of plant nutrients by *Euryale ferox* Salisb. growing in field condition in northern Bihar of north India. *Int. J. Curr. Microbiol. App. Sci.*, 2017, *6*, 1229-1237.

24. Kumar, A. and Jena, H. M. High surface area micro porous activated carbons prepared from Fox nut (*Euryale ferox*) shell by zinc chloride activation. *Applied Surface Science.* 2015 356: 753-761.

25. Kumar, L.; Singh, A.K.; Bhatt, B.P. Nutritional status of recently developed Makhana (Gorgon Nut) variety-Swarna Vaidehi. *J. Agri. Search*, 2016, *3*,199-205.

26. Kumar, L.; Gupta, V.K.; Jha, B.K.; Singh, I.S.; Bhatt, B.P.; Singh, A.K. Status of Makhana (*Euryale ferox* Salisb.) cultivation in India. Tech. *Bulletin*, 2011a, ICAR-RCER, Patna.

27. Kumar, L.; Gupta, V.K.; Khan, M.A.; Singh, S.S.; Jee, J.; Kumar, A. Field based Makhana cultivation for improving cropping intensity of rice fields. *Bihar J. Horti.*, 2011b, *1*, 71-72.

28. Li, L.J.; Wu, Y.F.; Cao, B.S. Research progress of *E. ferox* seeds. *China Vegetables*, 2007, 81-88.

29. Liu, X.; He, Z.; Yin, Y.; Xu, X.; Wu, W.; Li, L. Transcriptome sequencing and analysis during seed growth and development in *Euryale ferox* Salisb. *BMC Genomics*, 2018, *19*, 343.

30. Liang, K.; Liang, L. https://www.google.com/patents/ US20020031559 US Patent 2002, 'Herbal suppositories'.

31. Nath, B.K.; Chakraborty, A.K. Studies on the amino acid composition of the seeds of *Euryale ferox* Salisb. *J. Food Sci. Technol.*, 1985a, *22*, 293.

32. Negi, J. S.; Jugran, V. and Kasliwal, N. Development of non-effervescent floating matrix tablets based on *E. ferox* seeds. Asian Journal of Pharmacentics. 2011, 5(2): 93-100.

33. Puri, A.; Sahai, R.; Kiran, L.; Singh, R.P.; Saxena, K.C. Immunostimulant activity of dry fruits and plant materials used in Indian traditional medicinal systems for mothers after child birth and invalids. *J. Ethnopharmacol.*, 2000, *71*(1-2), 89-92.

34. Song, J.; Wu, Q.N. Analysis of therapeutic effects of Qian Shi on proteinuria. *China J. Mod. Drug Appl.*, 2009, 3, 133-134.

35. Sundaram, P.K.; Sarkar, B.; Mondal, S. Design and performance evaluation of pedal operated Makhana (*Euryale ferox* Salisb.) seed grader. *Res. J. Agri Sci.*, 2014, 5, 428-431.

36. Singh, I.S.; Kumar, L.; Gupta, V. K.; Bhatt, B.P.; Singh, S.P. Assessment of soil nutrient availability and productivity of Makhana (*Euryale ferox* Salisb.) based cropping systems in inceptisols of tropical humid climate of India. *Bioglobia*, 2014, *1*, 64-74.

37. Shen, B.; Qinan, W.U.; Rong, C.; Wei, Y.U.E.; Xiaoxiao, H.E. The modern research progress of *Euryale ferox. Northwest Provincial J.* (abstract) 2012-02.

38. Shoaib, S.M.; Khanam, D.S.; Jha, V.; Kumar, R.J.; Ahmad, A. Studies on dielectric properties of Gorgon nut [*Euryale ferox* Salisb.] *Bioglobia*, 2016, *3*, 35-44.

39. Uma Shankar, M.S.; Sateesh Madhev, N.V. Isolation and characterization of a novel smart mucoadhesive bioadhesive biopolymer from *Euryale ferox* seed coat. Pharmabit, 2011, *23-24*(1-2), 57-60.

40. Verma, S. K. Fishing Methods of Darbhanga district. M. Sc. Zoology Dissertation. L. N. Mithila University, Darbhanga, 1978.

41. Wu, C.; Wang, X.; Wang, H.; Shen, B.; He, X.; Gu, W.; Wu. Q. Extraction optimization, isolation, preliminary structural characterization and antioxidant activities of the cell wall polysaccharides in the petioles and pedicels of Chinese herbal medicine Qian *(Euryale ferox* Salisb.). *Int. J. Biol. Macromol.*, 2014, *64*, 458-467.

42. Wu, C.; Chen, R.; Wang, XS.; Shen, B.; Yue, W.; Wu, Q. Antioxidant and anti-fatigue activities of phenolic extract from the seed coat of *Euryale ferox* Salisb. and identification of three phenolic compounds by LC-ESI-MS/MS. *Molecules*, 2013, *18*, 11003-11021.

43. Yu, H.; Fang, J.; Zhang, X.; Sun, Y. Effect of Yishen capsule on podocytes in renal tissue of rats with diabetic nephropathy. *Chin. J. Integ. Trad. Western Nephrol.*, 2009, *10*, 577-579.

44. Yuan, H.; Meng, S.; Wang, G.; Gong, Z.; Sun, W.; He, G. Hypoglycemic effect of triterpenoid-rich extracts from *Euryale ferox* shell on normal and streptozotocin-diabetic mice. *Pak. J. Pharm. Sci.*, 2014, *27*, 859-864.

45. Zhang, R.; Cui, Z.M. Advance of *E. ferox* Salisb. fundamental and application research. *Chin. Sci. Technol. Inf.*, 2009, *26*, 130-132.

16

Documentation and Conservation Strategy Applied to Some Rare Plants of Pharkiya Region of Koshi Belt (Bihar)

Swati Sen and Umesh Kumar

[1] *University Dept. of Botany, B.N. Mandal University, Madhepura, India*
[2] *Dept .of Botany, Rosera College, Rosera Samastipur, India*

INTRODUCTION

This is by and large an ethnobotanical study involving two tuber bearing sedges called *Cyperus palustris* (Var. Chichor) and *Scirpusgrossus* (Var. Kysoor). These two plant species have been recognised as the 'Famine Food' and well documented in "A statistical Account of Bengal, Vol. IV by W.W. Hunter (1877) and Bihar & Orissa District Gazetteers (Monghyr) authored by L.S.S. O'Malley, 1925". But unfortunately, very scanty information on these two plant species is available despite the fact that they are vitally important for the survival of marginalised community of Musahars' or rat catchers, and that too in the seasons or at the time of scarcity. In the words of Lackwood: "When I saw 500 persons chichor hunting in a single marsh, I knew that there was famine in the land". According to W.W. Hunter: "A man may collect about a six pound a day, with average luck and labour, and they are usually ground into a kind of flour and made into bread or cakes, but are also eaten raw". These sedges have been rightly described as "Famine Thermometers". A party was digging up the underground creeping stem of water lily or the sedge bulbs called 'Chichor' in the vernacular, which are eaten to give flavour to the snails and crabs meat.

According to Vartak (1981) the tubers of *Scirpusgrossus* are sweet, starchy, highly nutritious and also eaten uncooked by some tribals in India. The crops have also found a mention as an Ethnobiological, information in the Arthasashtra of Kautilya.

OUR EFFORTS

With a combined effort of researchers, teams have been visiting the areas where endemic germplasms are found. These areas cover parts of the districts of

Khagaria and Saharsa which are characterised by ever changing topography and land profile due to the presence of many perennial rivers. Some of these rivers shift their banks very swiftly, cutting the margins and ultimately inundating their areas across. The soil composition is sandy, sometimes loamy and almost persistent annual floods leave cover of fertile silts behind.

Geographically the area is tough terrain with a comparatively thin human habitation and mostly occupied by socio-economically backward ethnic groups. The region as a whole is called "Pharkiya region". We have been able to collect and study the present growth behaviour and utility of certain indigenous species. During our course we finally selected two germplasms for our study, viz., *Cyperus palustris* and *Scirpusgrossus.*

THE HISTORY OF PHARKIYA REGION

Historically, it is said that during the times of Emperor Akbar, Raja Todarmal had been entrusted with the duty of making the survey of the entire area but he failed to do so because of the thatching grasses and shifting of the path of rivers. He advised that this area should be excluded. In other words, he adopted the policy of "Farak-Kiya" (to separate) and that is why the area since then is known as "Pharkiya Pargana or Belt or Region locally".

Geographically, this region has saucer-shaped depression and flooding of the area during rainy season is an annual feature on account of outburst of water from the seven major rivers- The Ganga, BurhiGandak, Bagmati, Kamla, Koshi, Kareh and Ghaghri.

This Pharkiya region is situated between latitudes 25°15'- 25°50'N and longitude between 86°20'-86°54'E. This region is characterised by alluvial soil. The average maximum and minimum temperature lies between 36.6°C and 20.9°C. The atmosphere is very humid in rainy season and slightly lesser in October and in winter months. The air is dried in summer season. The Pharkiya region is spread over an area of 752 sq. miles.

From ethnobotanical point of view, we have targeted extremely poor ethnic community- 'The Musahars', who are enlarged as landless labourers community. The study was done to examine the impact of selected germplasms on their livelihood and other consequential influences.

THE MUSAHARS

According to Hunter (1877), "The Musahars, or Mousers, as we should call them, catch and eat the field rat". Rat meat and rice comprise their food in the season of availability. As observed by O'Malley (1925), "The Musahars are field labourers whose wages are customarily paid in kind in the villages. They live in a kind of social thraldom, sometimes selling themselves, their wives and children to lifelong servitudes for paltry sums. With an ingrained aversion to emigration, pilfering in times of plenty, and living upon roots, rats, snails and shells, they cause considerable difficulty to Government officials in times of death. They live away from basti. The bulk of the castes are field labourers and palankeen bearers, and only a few have attained the dignity of cultivating on their own holdings or acquired occupancy rights. The name Musahar is believed to mean rat-catchers or rat-eaters and is an appropriate usage, for they are professional rat catchers.

Lockwood, in his work *"Natural History, Sport and Travel"*, writes: "The Musahars, or Mousers, as they may be called, are found in every village of Monghyr, and are half starved even in times of plenty. They seldom see coin, but receive their scanty wages in terms of grain, which they flavour with rats, mice, snails, and jungle roots, whilst living in hovels which an English Pig would consider poor accommodation. One would imagine that such persons must find difficulty in getting wives, but contrary is the case, for bachelors and spinsters are unknown. At the age of puberty they present themselves at their landlords house and having signed a deed binding themselves to remain in bondage for the terms of their natural lives, receive a few shillings in return, with which to entertain their friends at a marriage feast and set up house."

THE GERMPLASM

Cyperus palustris (Chichor) and ***Scirpusgrossus*** (Kysoor) are sedges belonging to family Cyperaceae.

1. *Cyperus palustris* (Chichor)

- Full grown plant reaches the height of approximately 1m and gives tufted look.
- Leaves are cylindrical, hollow, grooved and green with parallel venation.

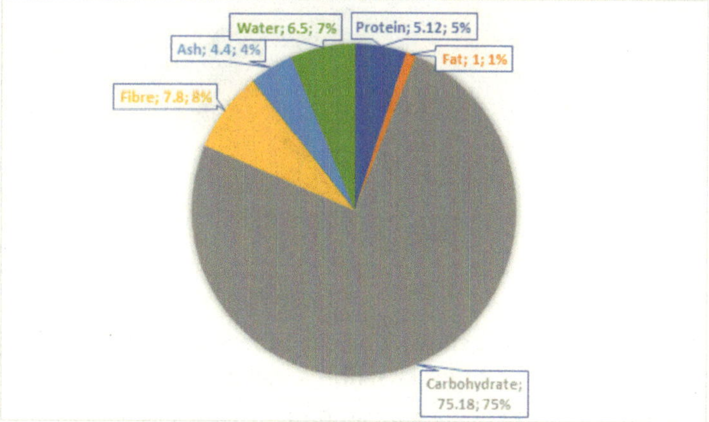

- Large-sized blackish brown, amorphous, oval, rhizomatous tubers are produced near the base of the underground pat.
- The mean length of tuber is 3.52 cm and average diameter has been measured to be 2.25 cm.

- The range of weight of individual tuber varies from 0.5-2g.
- The chemical composition of tubers in 100 gm is presented in the Pie chart:-
- A perusal of the chart shows that tubers are rich in carbohydrates in addition to some proteins, fats, fibres, ash and water.
- The calorie value of these food stuffs has been measured is 340 calories.

2. *Scirpusgrossus* (Kysoor)

- Fully grown plant reaches height of approximately 2 meters and gives tufted look.
- Stem is triangular and fleshy.
- Leaf is fleshy and 2cm wide in the middle. Parallel venation is present.
- Spikelet inflorescence with glumes present.
- Tubers look black and oval.
- The mean length of tuber is 7.12cm and average diameter measures 3.45cm.
- Weight of individual tuber varies from 6.5-8 gm.
- The chemical composition of tubers in 100gm is presented in the Pie chart (after Paton and Dunlop):-

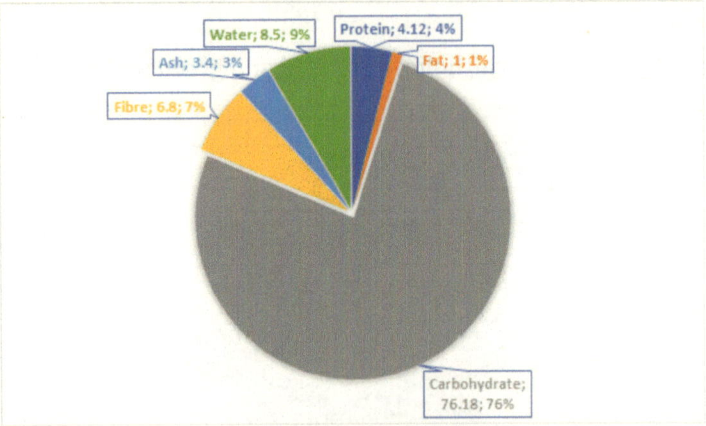

- A perusal of the chart shows that the tubers are rich in carbohydrates in addition to some proteins, fats, fibres, ash and water.
- The calorie value of these food stuffs, has been 338 calories.

UTILITARIAN VALUE OF GERMPLASM

As perceived earlier tubers of the endemic plants constitute a vital source of food for human survival in the periods of acute scarcity and therefore, they have rightly been called famine food. Apart from food, other plant parts like stems and leaves of these species are variously used, namely as fodder and for making household articles like mats, baskets and door curtains.

The collected germplasms have been conserved in the selected plots of the botanical garden. The cytological studies of these germplasms are under progress.

APPENDICES

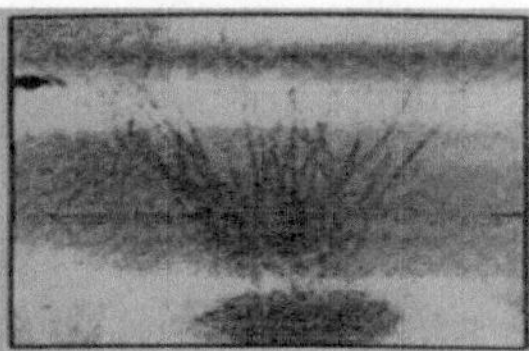

Fig. 1 (a) A tuft of chichor

Fig. 1 (b) A local boy holding tuft of Kysoor

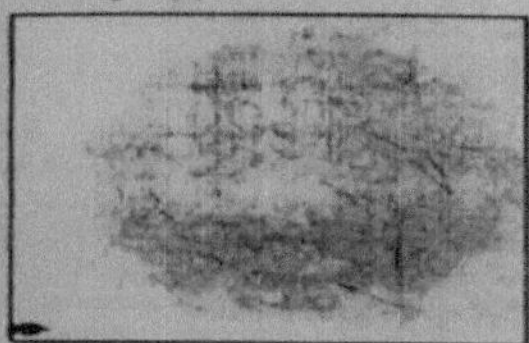

Fig. 2 (a) Tubers of chichor

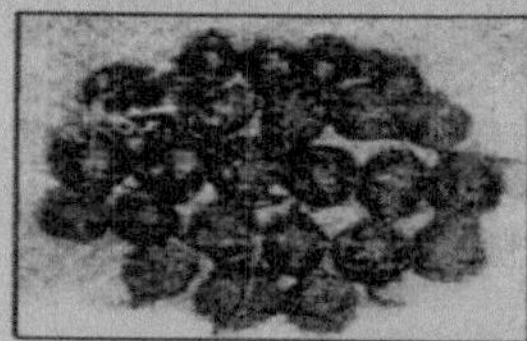

Fig. 2 (b) Tubers of Kysoor

Fig. 3 (a) Germinating Tubers of chichor

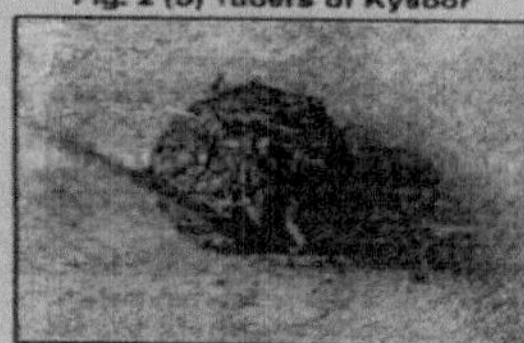

Fig. 2 (b) Germinating Tuber of Kysoor

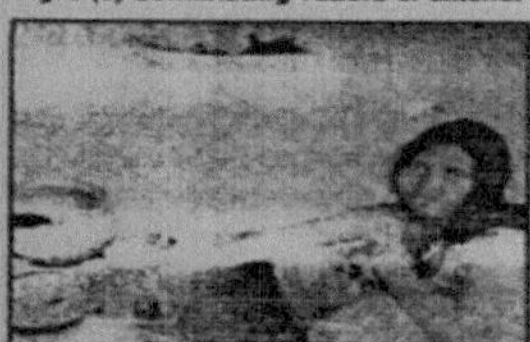

Fig. 4 (a) A native woman baking bread from flour of chicor

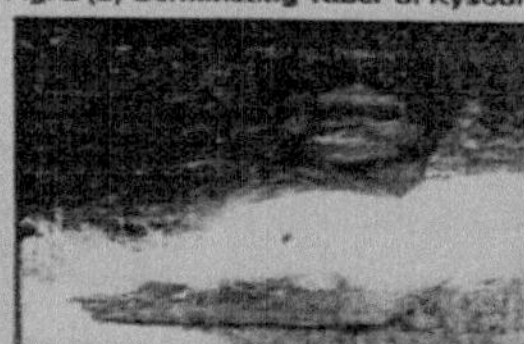

Fig. 4 (b) A native woman weaving mat from plants of kysoor

REFERENCES

1. Chandel, S.C.R, R.P. Upadhyay (2013). A survey of Ethnoflora of Pharkiya region of Koshi belt. Int. J.Mendel, 30 (1-4), 33-38.

2. Vartak. V.D., (1981). Observation on wild edible plants from hilly regions of Maharashtra and Goa: Resume and future prospects, in S.K. Jain (ed), Glimpses of Indian Ethnobotany. Oxford & IBH Publishing Co., New Delhi.

3. Rao AS, Verma DM (1982). Cyperaceae of North East India. Botanical Survey of India, P.O. Botanic Garden, Howrah, 93 pp. [+ VI.].

4. Dai L, Liang S, Zhang S, Tang Y, Koyama T, Tucker GC, Simpson DA, Noltie HJ, Strong MT, Bruhl JJ, Wilson KL, Muasya AM (2010). Cyperaceae. In: Wu ZY, Raven PH, Hong DY (Eds) Flora of China (Vol. 23) (Acoraceae through

Cyperaceae). Science Press, Beijing and Missouri Botanical Garden Press, St. Louis, 164–461.

5. Dey S, Prasanna PV (2015). Cyperaceae: Mapanioideae; Cyperaceae: Cyperoideae, Tribe Schoeneae&Sclerieae. In: Singh P, Dey S (Eds) Fascicles of Flora of India, 27. Botanical Survey of India, Kolkata, 118 pp.
6. Shrestha KK, Bhattarai S, Bhandari P (2018). Cyperaceae. Handbook of Flowering Plants of Nepal 1. Scientific Publishers, Jodhpur, India, 194–222.
7. Simpson DA (2019. Cyperaceae. In: Middleton DJ, Leong-Škornicková J, Lindsay S (Eds) Flora of Singapore (Vol. 7). National Parks Board Singapore, 37–211.
8. Govaerts R, Simpson DA, Goetghebeur P, Wilson KL, Egorova T, Bruhl JL (2021). World Checklist of Cyperaceae. Kew: The Board of Trustees of the Royal Botanic Gardens, Kew.

17

The Cyanobacterial Genus *Spirulina* from the Supaul District of North Bihar, India

Deepak Kumar Jha

Phycology Lab, University Dept. of Botany, L. N. Mithila University, Darbhanga, Bihar 846004, India

INTRODUCTION

Cyanobacteria, formerly known as blue-green algae, are a diverse group of Gram-negative organisms that originated around 3.5 billion years ago. Since then, they have contributed significantly through their photosynthetic activity, nitrogen-fixing abilities, and production of various valuable biomolecules. *Spirulina*, a filamentous edible cyanobacterium belonging to the family Oscillatoriaceae, has a distinctive spiral-shaped filamentous thallus with a characteristic blue-green color. It has gained global attention as a potential food source due to its high nutritional value. In addition to its high protein content, *Spirulina* is a rich source of vitamins, polyunsaturated fatty acids, and pigments such as phycocyanin, β-carotene, and chlorophyll. These bioactive compounds have been extensively utilized as natural colorants in the food and beverage industry, as well as in cosmetic and pharmaceutical applications.(Belay, 2008; Kamble and Karande, 2018).

The perusal of literature reveals that significant qunantity of research has been conducted on the presence and distribution of *Spirulina* (Nostocales) in India, as documented in various studies (Sinha and Srivastva, 1980; Jha *et al.*,2007; Rai *et al.*,2008; Nandan and Ahuja, 2010; Ansari *et al.*, 2012; Kumar and Sahu, 2012; Roy and Keshri, 2014; Kamble and Karande, 2018 etc.). However, the information of the taxonomy of this order from Bihar is scanty and very sporadic (Sinha and Srivastava, 1980; Mohan and Kumar, 2019). Thus, our knowledge about the order Nostocales is poorly understood from North Bihar. Therefore, considering the importance of the algal research on this order, the present taxonomic study has been taken up. The present communication is the results of algal collections made from different localities of Supaul district of North Bihar.

MATERIALS AND METHODS

Field Survey: Algal samples were gathered during January 2021–January 2023 from a variety of permanent and temporary freshwater habitats of Supaul district (25°37' - 26°25'N Latitude and 86°22'-87°10'E Longitude), northern Bihar. These samples were carefully collected following the standard procedure, using small plastic containers. Each sample was assigned a 'Collection code' based on the location of its collection. The samples were preserved in FAA with Glycerine (50 ml per liter of the preservative). The pH and temperate of the water body at collection sites were recorded.

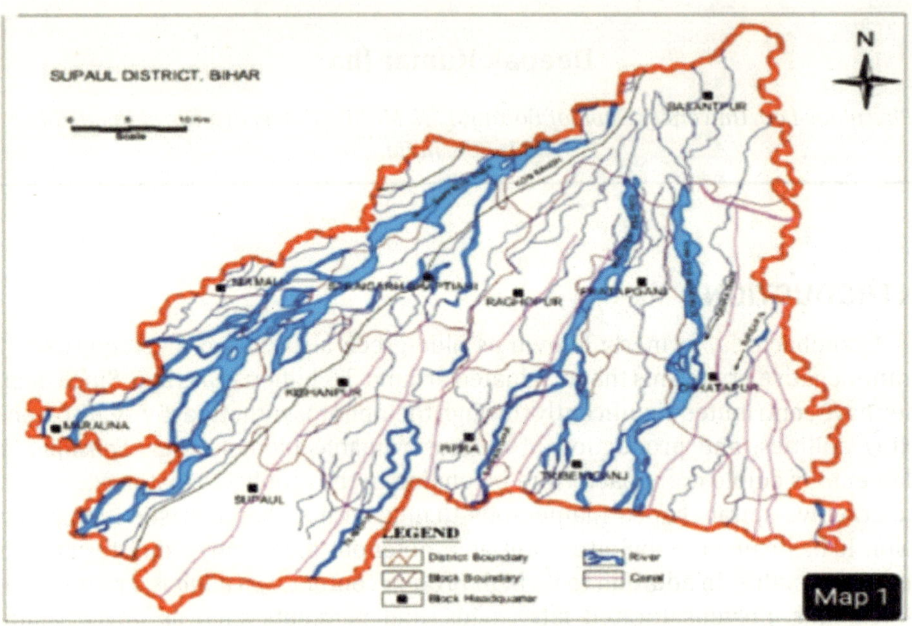

Figure 1: Map of Supaul District

Laboratory study: The samples were observed under Olympus 7213 and Prism type Camera Lucida was used for drawing. Microphotographs of the material were captured by Olympus digital camera (Model E-420). The voucher specimens are kept in the Phycology Laboratory, University Department of Botany, L.N. Mithila University, Darbhanga, Bihar.

Identification: The identity and distribution of the collected samples were confirmed following the standard literature (Desikachary, 1959; Roy and Keshri, 2014 ; Mohan and Kumar, 2019 etc.).

RESULTS

In the present investigation 2 species of *Spirulina* i.e., *S. princeps* West & G.S. West and *S. gigantea* Schimidle are reported.

1.*Spirulina princeps* West & G.S. West (Plate 1: Fig.A)

(Desikachary, 1959; P.197; Pl.36; Fig.7)

Thallus blue green and filamentous, trichome 4.5- 5.6μm broad, 80-186 μm long, regularly spirally coiled, distance between spirals 7-9.5μm.

The present specimen is similar to the type species.

Habitat: Collection No. DK-501; Date- January 21, 2023; collected from a road side canal(pH 6.5, Temp.26°c) of Pilwaha, Tribeniganj (Dist. Supaul) along with *Pediastrum* species.

Distribution: Karnataka, Assam, Bihar, Chattisgarh, Assam, Bihar, Chattisgarh, Gujarat, Jammu & Kashmir, Kerala, Madhya Pradesh, Maharashtra, Odisha, Punjab, Rajasthan, Tamil Nadu, Uttarakhand, Uttar Pradesh, West Bengal(in FLORA OF KARNATAKA- A Checklist, Karnataka Biodiversity Bord 2019)

2.*Spirulina gigantea* Schimidle (Plate 1: Fig.B)

(Desikachary, 1959, p. 197, pl. 36, figs.12, 14,-17)

Thallus deep blue green and filamentous, trichome 3.7-4.2 μm broad, regularly spirally coiled, at the end conical attenuated, spirals 10-16.5 μm broad.

Habitat: Collection No.DK-18, Date- April 21, 2021, from a ditch (pH 6.0, Temp. 32 °C) of Basantpur (Dist. Supaul) growing along with *Anabaena* and *Spirogyra* species.

Distribution: Karnataka (Mysuru), Assam, Chattishgarh, Gujarat, Jammu & Kashmir, Kerala, Maharashtra, Meghalaya, Odisha, Punjab, Tamil Nadu, Uttarakhand, Uttar Pradesh, West Bengal.(in FLORA OF KARNATAKA- A Checklist, Karnataka Biodiversity Bord 2019)

This is the first record of the taxon from Bihar.

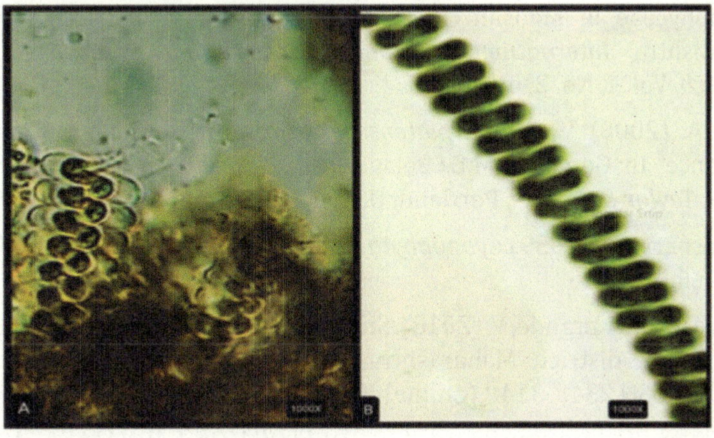

Plate 1: Fig.A: *Spirulina princeps* West & G.S. West ; Fig.B: *Spirulina gigantea* Schimidle

RESULTS AND DISCUSSION

Previous research on cyanobacterial species, particularly within the order Nostocales, has been sparse in Bihar, as highlighted by Sinha and Srivastava (1980) and Mohan and Kumar (2019). The identification of *Spirulina* species in this study aligns with findings from other regions of India, where these taxa have been reported in a variety of habitats, including ditches, ponds, rice fields and roadside canals (Desikachary, 1959; Kumar and Sahu, 2012; Kamble and Karande, 2018). However, their presence in the unique ecological conditions of North Bihar offers new insights into the adaptability and ecological significance of these cyanobacteria.

Conclusion

The present investigation successfully records two species of the cyanobacterial genus *Spirulina*, namely *S. princeps* and *S. gigantea*, from Supaul district, North Bihar. This study contributes significantly to the knowledge of cyanobacterial diversity in the region, particularly regarding the order Nostocales, which has been poorly documented in previous studies. The findings not only expand the known distribution of *Spirulina* species but also highlight their ecological significance and potential for various commercial applications due to their high nutritional value. The discovery of *Spirulina gigantea* as a new record for Bihar is particularly noteworthy, further enriching the algal biodiversity of the state. The results underscore the need for continued taxonomic exploration and documentation of cyanobacteria in North Bihar, which could lead to the discovery of more novel species and provide a basis for future research on their ecological roles and biotechnological potential.

REFERENCES

1. Ansari, Z., Tambe, S. S., and Nandan, S.N. (2012), "Biodiversity studies of Cyanophyceae in Mausam River of Malegaon city, district: Nasik, state: Maharashtra, *International Journal of Science and Pharma Educational Research*, Vol. 1, No. 2, pp. 69-72.

2. Belay, A. (2008) "*Spirulina platensis* (*Arthrospira*): production and quality assurance". In: Gershwin M.E.; Belay A. (Ed.). *Spirulina* in human nutrition and health. *Taylor & Francis*, Portland, USA, pp.2-23.

3. Desikachary, T.V.(1959).*Cyanophyta*, IndianCouncil of Agricultural Research, New Delhi.

4. Kamble, P. and Karande, V. (2018). Studies on cyanobacteria: *Spirulina* isolated from Satara district, Maharashtra. *International Journal of Life Sciences Research* ISSN 2348-3148 (online) Vol. 6, Issue 3, pp: 261-270

5. Karnataka Biodiversity Board, (2019). *FLORA OF KARNATAKA, A Checklist. Volume – 1: Algae, Fungi, Lichens, Bryophytes & Pteridophytes* . 1-562 (Published by Karnataka Biodiversity Board)

6. Kumar, A. and Sahu, R. (2012), "Ecological studies of Cyanobacteria in sewage pond of H.E.C. industrial area, Ranchi India", *Bioscience Discovery*, Vol. 3, No. 1, pp. 73–78.

7. Mohan, A., and Kumar, B.(2019). Cyanobacterial diversity in agriculturally fields soil of Patna and their population density. *ISOR.*5(1):15-32

8. Nandan, S. N. and Ahuja, S. R. (2010), "Study of blue green algal biodiversity of lentic hydrosphere of Haranbari dam of Maharashtra (India)", *Asian J. Exp. Biol. Sci.* (Spl.), pp. 132–135.

9. Rao, U.N., Dubey, S., Shukla, O.P., Dwivedi, S. and Tripathi, R.D. (2008), "Screening and identification of early warming algal species for metal contamination in fresh water bodies polluted from point and non point sources", *Environ. Monit. Assess.*, Vol. 144, pp. 469-481.

10. Roy, S. and Keshri, J.P. (2014). On the occurence of the members of Nostocales (Cyanophyta) from Burdman, West Bengal, India with a note on their ecology. *Int. J. LifeSc. Bt & Pharm. Res.* Vol. 3, No.3 , p.126-149.

11. Sinha, B.D. and Srivastava, N. K. (1980), "Algal flora of Balmikinagar, Bihar", *Phykos*, Vol. 19, pp. 171-174.

18

Biochemical Characterization of Pathogen Causing Crown Rot of Papaya in Samastipur

Ajeet Kumar*[1] and S.N. Jha[2]

[1]University Dept. of Botany,
L.N. Mithila University, Darbhanga, Bihar, 846004.
[2]K.S.R. College, Sarairanjan, Samastipur
Email: ajeetkumar1301@gmail.com

INTRODUCTION

Papaya (*Carica papaya* L.) is a fruit of tropical and subtropical regions, useful for its nutritional, medicinal and commercial benefits. In India, papaya is an important horticultural crop, and Bihar is one of the major states contributing to its production. However, papaya cultivation faces many challenges from diseases. Crown rot, a new outbreak of papaya disease affecting the plants, particularly observed in Samastipur. The causal pathogens have been linked to species such as Erwinia spp. Studies have identified *Erwinia mallotivora* as a primary agent in other countries like Indonesia and Malaysia (Suharjo *et al.,* 2021).

In Indonesia, the 1st crown rot disease of papaya was reported by von Rant (1931) at Java Island, the pathogen was known as *Bacillus papayae*. The bacteria were placed within genus Erwinia by Magrou (1937). Later, Gardan et al. (2004) proposed a new name of the pathogen as *Erwinia papayae*. This pathogen has been reported causing papaya bacterial canker in the Caribbean islands (Gardan et al. 2004).

There is limited research on the biochemical properties of these pathogens in Bihar, India. Despite the incidence of crown rot in Samastipur, there is a lack of biochemical and molecular characterization of the responsible pathogens, impeding effective management strategies. Crown rot is a bacterial disease that leads to extensive damage, reducing yield and economic viability for farmers. This disease causes wilting and rotting of the crown area, which, if left untreated, leads to the death of the plant.

Effective disease management strategies depend on the accurate identification of the causal pathogen. Biochemical assays, such as those described in Holt *et al.*, Bergey's Manual of Determinative Bacteriology, offer reliable means to identify bacterial species based on morphological, physiological, and biochemical characteristics (Suharjo *et al.*, 2021).

The identification of *Erwinia mallotivora* or related bacterial species in Samastipur can help develop targeted control measures, thus reducing crop losses. Characterizing the pathogen through biochemical assays and understanding its growth behaviour at different environmental conditions are crucial steps in this process.

MATERIALS AND METHODS

The study was conducted in Samastipur, Bihar, where papaya (*Carica papaya* L.) plants exhibiting symptoms of bacterial crown rot were sampled. A total of 15 symptomatic plants, including those showing wilting, decay, and necrosis, were collected from five major papaya-growing areas within the district. Samples were taken from the roots of affected plants and transported to the laboratory in sterile bags, stored at 4°C, and processed within 24 hours (Sharma *et al.*, 2019).

The isolation of the bacterial pathogen was performed by surface sterilizing the infected root tissues using 70% ethanol for 2 minutes, followed by three rinses in sterile distilled water. Small sections of sterilized tissue were macerated and plated onto nutrient agar (NA) medium using the streak plate method (Holt *et al.*, 1994). Plates were incubated at 28°C for 48 hours, and the developing colonies were subcultured to obtain pure isolates. Morphological observations of colonies, including shape, size, colour, and texture, were recorded to preliminarily characterize the pathogen (Kumar *et al.*, 2020).

Biochemical Assays

To further characterize the isolated bacterial pathogen, a series of biochemical assays were conducted following standard procedures outlined by Holt *et al.* (1994) in Bergey's Manual of Determinative Bacteriology. The biochemical tests included:

Gram Staining, procedure was followed to determine the Gram reaction (Cappuccino & Sherman, 2019). A loopful of the bacterial isolate was smeared on a glass slide, heat-fixed, and stained using crystal violet, iodine, alcohol, and safranin. The bacterial cells were observed under a light microscope to determine whether they were Gram-positive (purple) or Gram-negative (pink).

KOH String Test, to confirm the Gram reaction, a 3% potassium hydroxide (KOH) solution was used. A loopful of bacterial culture was mixed with a drop of KOH on a glass slide and stirred for 10 seconds. The formation of a viscous string indicated a Gram-negative bacterium (Halebian *et al.*, 1981).

Catalase Test, was assessed by placing a drop of 3% hydrogen peroxide on a glass slide and mixing it with a loopful of the bacterial colony. The production of oxygen bubbles indicated a positive result, confirming the presence of catalase enzyme (Holt *et al.*, 1994).

Oxidase Test, was tested by adding the bacterial colony to a filter paper soaked in 1% tetramethyl-p-phenylenediamine dihydrochloride solution. A colour change to purple within 20 seconds indicated a positive result, showing the presence of cytochrome oxidase (Kovács, 1956).

Urease Test, used to know the ability of the bacterial isolate to hydrolyze urea using Christensen's urea agar slants. The inoculated slants were incubated at 28°C, and the production of a pink colour after 24 hours indicated urease activity (Cappuccino & Sherman, 2019).

Gelatin Liquefaction Test, was conducted by inoculating the bacterial isolate into nutrient gelatin deep tubes. After incubation at 28°C for 5-7 days, liquefaction of the medium indicated the presence of gelatinase enzyme (Holt *et al.*, 1994).

Koch's Postulates

The suspected causal agent is isolated and reinoculated into a healthy plant the host must reproduce same disease that was seen earlier (Agrios, G. N., 2004).

To confirm the pathogenicity of the isolated bacterium, Koch's postulates were applied. Healthy papaya seedlings were grown in sterile soil and inoculated with a bacterial suspension (10^8 CFU/mL) of the isolate. The bacterial suspension was prepared by growing the bacterial culture in nutrient broth for 24 hours. The inoculation was performed by drenching the soil around the roots of the plants with 20 mL of the bacterial suspension (Lelliott & Stead, 1987). A control group was treated with sterile water.

The plants were monitored for the development of root rot symptoms over a period of 14 days. Once symptoms appeared, the pathogen was re-isolated from the infected roots using the same method described earlier. The re-isolated bacterium was compared to the original isolate in terms of morphology and biochemical characteristics, fulfilling Koch's postulates (Lelliott & Stead, 1987).

Growth at Different Temperatures

The growth of the bacterial isolate was assessed at different temperatures to determine its optimum growth conditions. The bacterium was inoculated into nutrient broth and incubated at 20°C, 28°C, and 37°C. Growth was measured by monitoring optical density (OD600) at 24-hour intervals for 72 hours using a spectrophotometer (Kumar *et al.*, 2020). The results were plotted to determine the temperature at which maximum growth occurred.

RESULTS AND DISCUSSION

This section presents the findings from bacterial isolation, biochemical assays, growth at different temperatures, the pathogenicity test and identification of pathogen.

Bacterial Isolation

From the symptomatic papaya samples collected from Samastipur, bacterial colonies were successfully isolated. The colonies appeared on nutrient agar (NA) plates within 48 hours of incubation. The colonies were shiny, and creamy-white, and after 72 hours (Fig 2), some turned purple on potato peptone glucose agar (PPGA) medium, which is characteristic of Erwinia spp. These isolates were preserved on PPGA slants for further biochemical and pathogenicity tests.

Fig 1. The Crown Rot symptom observed water-soaked lesions on stem damage caused by the disease, causing the death of papaya plant.

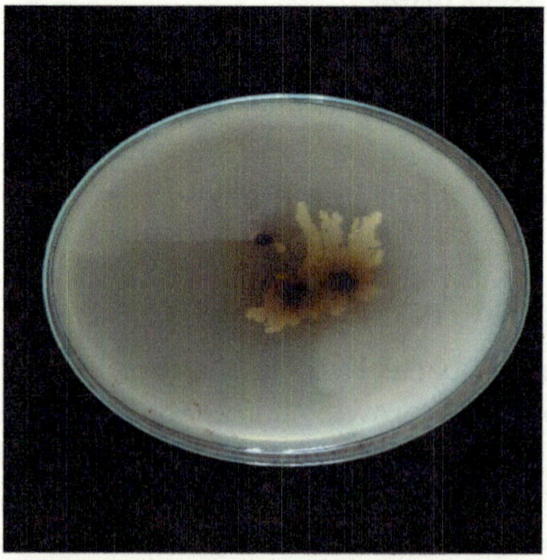

Fig 2. Bacterial colonies observed shiny, and creamy-white, and after 72 hours of inoculation in Nutrient Agar.

Biochemical Assays

The biochemical assays were performed to determine the characteristics of the isolated bacterial strains. The results are summarized below:

All the isolates were Gram-negative, confirmed by the viscous string formation in the KOH test. The bacteria showed fermentative metabolism, as indicated by acid production in the OF test. The isolates were catalase-positive, producing bubbles upon the addition of hydrogen peroxide, indicating the presence of the enzyme. All isolates were oxidase-negative. One isolate, designated AP07, was urease-positive, indicating its ability to hydrolyse urea into ammonia and carbon dioxide. This trait is commonly seen in *Burkholderia cepacia* but is rare in Erwinia. The isolates could utilize several organic compounds, including citrate, glycerol, and mannitol, as carbon sources. However, they could not utilize starch, lactose, or inulin.

Table 1. Biochemical Analysis of Pathogen causing Crown Rot of Papaya

Isolate code	Gram Reaction	Shape	Catalase	Oxidase	Urease	Indole	KOH String	Gelatin Liq.
	-	Rod	+	-	-	-	-	-
	-	Rod	+	-	-	+	-	-
	-	Rod	+	-	-	-	-	-
	-	Rod	+	-	-	-	-	-
	-	Rod	+	-	-	-	+	+
	-	Rod	+	-	-	-	+	+
	-	Rod	+	-	+	-	+	-

Growth at Different Temperatures

The growth of the bacterial isolates was tested at different temperatures (36°C, 37°C, 39°C, and 40°C). Optimal growth was observed at 36°C and 37°C, with slower growth at 39°C and 40°C. The bacteria failed to grow at temperatures above 40°C. These findings suggest that the pathogen thrives best in moderate to high temperatures, typical of the climatic conditions in Samastipur.

Pathogenicity Test (Koch's Postulates)

To confirm the role of the isolated bacteria as the causal agent of crown rot, Koch's postulates were followed:

Within 3-5 days of inoculation, the inoculated papaya plants began showing symptoms of water-soaked lesions on the petioles and leaves. By day 12, crown rot symptoms became evident, and by day 21, the plants exhibited complete wilting and death.

The bacteria were re-isolated from the diseased plants, and their biochemical profiles matched the original isolates, thus confirming that the isolated bacterium was the pathogen responsible for causing crown rot in papaya.

Identification of Pathogen

The results of this study confirm that the bacterial isolates obtained from symptomatic papaya plants in Samastipur are likely responsible for causing crown rot. The findings are consistent with those reported in other regions, particularly Southeast Asia, where similar bacterial species have been implicated in papaya diseases. This section will discuss the findings in relation to previous studies and highlight the significance of the results in advancing our understanding of papaya crown rot.

The isolated bacteria from Samastipur displayed characteristics that align closely with Erwinia spp., particularly *Erwinia mallotivora*. This pathogen has been widely reported as the causal agent of crown rot in papaya across Southeast Asia, including Malaysia and Indonesia. In Malaysia, *Erwinia mallotivora* was identified as the primary agent of papaya crown rot in 2011, where it caused significant losses in papaya production. Similarly, in Indonesia, *Erwinia mallotivora* was identified as the causative pathogen of crown rot in Lampung, displaying similar symptoms to those observed in our study.

The results of the biochemical assays in this study align with the findings of Suharjo *et al.* (2021), who reported that *Erwinia mallotivora* isolates were Gram-negative, catalase-positive, and fermentative. Like in our study, they also observed the inability of the bacteria to utilize certain carbon sources, such as starch and lactose, while showing the ability to metabolize citrate and mannitol. These shared biochemical traits further support the identification of the pathogen as *Erwinia mallotivora* in Samastipur.

Urease Activity

Interestingly, one of the isolates in our study, AP07, exhibited urease activity, a feature that is less commonly reported in Erwinia spp. but has been associated with other bacteria, such as *Burkholderia cepacia*. This observation differs from other studies, such as Gardan *et al.* (2004), which did not report urease-positive isolates of *Erwinia papayae*. This suggests that the pathogen population in Samastipur may include a diverse range of strains with varying metabolic capabilities, possibly due to environmental factors or local plant physiology.

Pathogenicity and Symptomatology

The successful completion of Koch's postulates in this study confirmed that the isolated bacteria were indeed responsible for the observed crown rot symptoms in papaya. The progression of symptoms, from water-soaked lesions to crown rot and eventual plant death, mirrors the disease development reported by previous studies on *Erwinia mallotivora* in Malaysia.

However, one notable difference is the time frame for symptom development. In our study, crown rot symptoms appeared within 12 days of inoculation, while Suharjo *et al.* (2021) reported symptom onset within 3-5 days. This variation could be due to differences in environmental conditions, such as temperature and humidity, which are known to influence the growth and virulence of bacterial pathogens.

Temperature and Growth

The growth experiments in this study demonstrated that the isolates grew optimally at 36°C and 37°C, with reduced growth at higher temperatures. These findings are consistent with previous research, which reported that *Erwinia mallotivora* thrives in moderate to high temperatures typical of tropical regions Suharjo *et al.* (2021). This temperature range corresponds to the climatic conditions in Samastipur, indicating that the pathogen is well-adapted to local environmental conditions.

In contrast, Gardan *et al.* (2004) reported slightly lower optimal growth temperatures for Erwinia papayae in the Caribbean, where temperatures are generally more temperate. The ability of *Erwinia mallotivora* to grow at higher temperatures may explain its dominance as a pathogen in tropical climates like Bihar.

CONCLUSIONS

The identification of *Erwinia mallotivora* as the causal agent of crown rot in Samastipur has important implications for disease management. Given its ability to thrive at higher temperatures and its aggressive pathogenicity, effective management strategies should focus on reducing environmental stress on papaya plants, particularly during the hot summer months.

Biological control options, such as the use of antagonistic microorganisms, could also be explored as a sustainable management approach. Studies on the biocontrol potential of bacteria like Burkholderia cepacia and fungi such as Trichoderma spp. have shown promise in managing bacterial diseases in papaya . These strategies could be integrated into current pest management programs to reduce reliance on chemical controls, which are often ineffective against bacterial pathogens like Erwinia.

Further research is needed to fully characterize the diversity of Erwinia species in the region and to investigate the potential presence of mixed infections with other bacterial species, such as Burkholderia. Molecular techniques, such as 16S rRNA sequencing, could provide more detailed insights into the genetic variability of the isolates from Samastipur, contributing to a deeper understanding of the epidemiology of crown rot in this region.

REFERENCES

1. Agrios, George N. (2004) *Plant Pathology*. Elsevier Science. 5th Edition
2. Gardan L, Christen R, Achouak W, Prior P. (2004) Erwinia papayae sp. nov., a pathogen of papaya (Carica papaya). *Int J Syst Evol Microbiol*. Jan;54 (Pt 1):107-113. doi: 10.1099/ijs.0.02718-0. PMID: 14742466.
3. Holt, J. G., Krieg, N. R., Sneath, P. H. A., Staley, J. T., & Williams, S. T. (1994). Bergey's Manual of Determinative Bacteriology. 9th Edition. Lippincott Williams & Wilkins.
4. Suharjo, R., Oktaviana, H. A., Aeny, T. N., Ginting, C., Wardhana, R. A., Nugroho, A., and Ratdiana, R. (2021). Erwinia mallotivora is the causal agent of papaya bacterial crown rot disease in Lampung Timur, Indonesia. *Plant Protection Science*, 57(2), 122–133. https://doi.org/10.17221/123/2020-PPS
5. Gardan, L., Christen, R., Achouak, W., and Prior, P. (2004). Erwinia papayae sp. nov., a pathogen of papaya (Carica papaya). *International Journal of Systematic and Evolutionary Microbiology*, 54(1), 107–113. https://doi.org/10.1099/ijs.0.02761-0
6. Magrou J. (1937): Dictionnaire des Bacteries Pathogenes pour l'Homme, les Animaux et les Plantes. In: Hauduroy P., Ehrdinger G., Urbain A., Guillot G., Magrou J. (eds): Paris, Masson & Cie.

19

Study of Invasive Wetland Plant Species of Madhubani, North Bihar

Ajay Kumar

Dept. of Botany, Parwati Science College,
B.N. Mandal University, Madhepura, Bihar

INTRODUCTION

Wetlands are a crucial natural resources. It is life support system that insure functions of water cycle ever since civilization began, wetland have played an important role in development of human society (Ansari and Kumar 2021). The wetlands represent a stable environment where living things interact and materials are used over and over again. In the past year, aquatic ecology has witnessed many alternations in its natural composition which is in general causes habitat distructions, fragmentation and loss of biological components Kumar (2018).

Invasive wetland plant species are non-native organisms that cause or have the potential to cause harm to the environment economies or human health. Shukla *et al.,* (2009) and Anonymous 2003. Aquatic ecosystem is of vital concern to humans and other living beings. In addition to livelihood option it helps the recharge of ground water, recycling of nutrients and purification of water. It provides habitats for a variety of flora Wilcoves *et al.,* (1998).

MATERIAL AND METHODS

In the present survey the authors documented 20 invasive plant species belongs to 21 genera and 17 families.

A list was prepared on the basis of recent collection and published literature. These invasive species along with family name, habit, habitat and nativity are listed in table 1.

Table-1: Invasive plant species of Rahika wetland, Madhubani, Bihar.

Sl. No.	Name of Plant	Family	Nativity	Habit	Habitat
1	*Alternanthera sessilis* (L.)	Amaranthaceae	Tropical America	Herb	Common in aquatic and semiaquatic
2	*Amaranthus spinosus* L.	Amaranthaceae	Tropical America	Herb	Wastelands
3	*Antigonon leptopus* Hook.	Polygonaceae	Tropical America	Climber	Along road sides
4	*Blainvillea acmella* (L.)	Asteraceae	Tropical America	Herb	Wastelands
5	*Blumea eriantha* DC.	Asteraceae	Tropical America	Herb	Wastelands
6	*Caldesia oligococca* (F.v.Muell.)	Alismataceae	Australia	Herb	Muddy places
7	*Cassia alata* L.	Caesalpiniaceae	West Indies	Shrub	Wastelands
8	*Ceratophyllum demersum* L.	Ceratophyllaceae	Tropical America	Herb	Aquatic
9	*Corchorus aestuans* L.	Tiliaceae	Tropical America	Herb	Wastelands
10	*Cyperus difformis* L.	Cyperaceae	Tropical America	Herb	Paddy fields
11	*Eichhornia crassipes* (Mart.)	Pontederiaceae	Tropical America	Herb	Ponds
12	*Eclipta prostrate* L.	Asteraceae	Tropical America	Herb	Wastelands
13	*Ipomoea carnea* Jacq.	Convolvulaceae	Tropical America	Shrub	Along border of Cultivated fields
14	*Lantana camara* L.	Verbenaceae	Tropical America	Shrub	Dry deciduous forests
15	*Ludwigia octovalvis* (Jacq.)	Onagraceae	Tropical America	Herb	Along edges of ponds
16	*Ocimum americanum* L.	Lamiaceae	Tropical America	Herb	Wastelands
17	*Parthenium hysterophorus* L.	Asteraceae	Tropical North America	Herb	Wastelands
18	*Ruellia tuberosa* L.	Acanthaceae	Tropical America	Herb	Moist shady places
19	*Solanum nigrum* L.	Solanaceae	Tropical America	Herb	Wastelands
20	*Typha angustifolia* L.	Typhaceae	Tropical America	Herb	Common in marshy and water logged places

RESULTS AND DISCUSSION

Invasive in the Barail wetland found in all kinds of habitats like forest, crop lands, wasteland, road side, pond banks and aquatic habitats maximum invasive species are found in wastelands, followed by cultivated land and pond bank. It is known that habitats differ in their invisibility. A habitat susceptible to invasion may possess the following attributes:- Species poverty, Poorly adapted native species, Absence of predators, Gaps generated by disturbance and Presence of empty niches Singh et al., (2001).

The region of north Bihar is very much in need of a proper "wetland management authority with a mandate of utilization of locally available natural resources vis-à-vis livelihood prospects in flooded area.

CONCLUSION

The present study reveals that there are 20 invasive species in the Barail wetland. Monitoring of invasion may be undertaken through qualitative approaches like species inventory and quantitative approach using phytosociological methods. It is essential to adopt strict measures at government and non-government levels including public awareness to control the populations of these invasive species, the places that are not yet invaded should be selected as priority areas for good management to prevent invasion.

REFERENCE

1. Anonymous. The IAS problem. The global Invasive Species Programme, (2003). http://www.gisp.org/about/IAS.asp.
2. Ansari, M.K. And Kumar, A. (2021). Physico-chemical observation of water in Rahika wetland Madhubani, District, Bihar. Souvenir international conference on innovative research in life science for sustainable development. 33-34.
3. Kumar, A, (2018). Water analysis of wetlands in Rahika Block of Madhubani District, Bihar. Academic, 6: 122-125.
4. Shukla, A.N., Singh K.P. and Singh J.S. (2009). Invasive alien species of Achanakmar-Amarkantak Biosphere Reserve, Central India. Proc. Nat. Acad. Sci. India, Sect. B, 79 (IV), 384-392.
5. Singh NP, Khanna KK, Mudgal V, Dixit RD. (2001). Flora of Madhya Pradesh. Vol.-3, Calcutta: Botanical Survey of India.
6. Wilcove DS, Rothstein D, Dubow J, Philips A, Losos E. (1998). Quantifying threats to imperilled species in the United States. BioScience, 48 : 607-615.

20

Makhana: Role in Food and Health Security

Ranjana

Dept. of Botany, D.B. College, Jaynagar,
Madhubani L.N. Mithila University, Darbhanga, 846004

INTRODUCTION

"Wetland is not wasteland. It is the kidney of the nature. They also pointed out their value of wildlife conservation, flood control, ground water supply, waste water treatment, fish production and Makhana cultivation. The processing of Makhana cro is only done by the fisher women.

Euryale ferox, commonly known as prickly, waterlily or Makhana is a species of waterlily, found in southern and eastern Asia and the only extant member of the genus Euryale. The edible seeds called fox nuts or Makhana. Makhana is a high value aquatic cash crop and a traditional Indian snack. It grows in stagnant perennial water bodies.

Makhana has been widely used in traditional oriental medicine to cure various diseases. They rich in nutrients and are a highly source of Manganese, Potassium, Thiamin,, Magnesium, Phosphorus and protein. In both raw and fried Makhana is rich in essential amino-acids. Slightly roasted Makhana are a perfect Tiffin. In India people make dishes such as Kheer, Raita, Curry and Cutlets using Makhana. In north Bihar people consume the seed in popped form. Other side of India people cook the leaves and stalk of the Makhana and enjoy it in vegetable curry. It is a perfect Tiffin option for kids. In India the state of Bihar is the largest producers of Makhana. Makhana are derived from lotus seeds. Lotus develops seeds pods and each pod contains approximately 20 seeds that matures within 40 day. The seeds are then dried and roasted on high flames. The outer black shell breaks and white puffs pop come out. These seeds are called Makhana. Makhana cultivation is also highly technical. It is a very valuable cash crop of Bihar. It is a priced delicacy with very high market demand. Bihar produces the 90% of world's Makhana. They made a farmer happy farmer. Makhana is also known for the anti-dialectic property. Research conducted at the Research Centre for Makhana Darbhanga found that Makhana based products are low in sugar content.

We prepared Makhana Barfi and Kalakand, Pakora as evening snacks, sweets and Makhana Chapati also make from Makhana flour and mixed flour. Makhana Kalakand was found to be low in sugar and high in protein as compared to Makhana Barfi's sugar and protein content. Makhana is a great snack and has multiple benefits. International stars eat them as midday or a post workout snack for that boost of energy. This has led to a sudden rise in the popularity of Makhana. It is also easily accessible and is a great snacking option. It is a great option for munching and maintaining an ideal weight.

They were extremely underrated earlier. It has gained new popularity in the last few years as people have become more health conscious.

Today veganism is a growing trend. A vegan diet is also called a plant based diet. Vegrism is a great choice vegans often fall short of their daily nutritional requirement. However, they are often deprived of the essential vitamins, minerals and protein, in their diet. This can lead to deficiency based health disorders in the long run. This is where Makhana can come and fill the need gap. They are loaded with Phosphorus, Zinc, Calcium, Protein, Magnesium and fibre etc. They are a perfect choice for a vegan lifestyle. It is a great vegan snack for those untimely hunger pangs. These low calorie nuts are a great option for guilt free binging.

They are rich in fibre which helps keep the excretory tract healthy. Makhana though small are a power house of nutrients. They can give amazing health benefits like controlled weight, glowing skin, good heart health, hormonal balance. 100 grams Makhana contains calorie-347, protein-9.7 gm, Fats-0.1gm, Carbohydrates-1.4 mg, File-14.5 gm, Total lipid-0.1 gm. They detoxify and cleans the spleen. Fox nuts help the Ever to function properly and enhance metabolism.

They help regulate blood sugar levels. They also help diabetic, patients to maintain their sugar levels. They are naturally bestowed with astringent properties which means people with weak kidneys can easily from Makhana. When we want to weight loss, it is beneficial to consume 30 gms of fox nut daily. The daily intake of foxnuts based on our specific needs. Excess consumption may lead to side effects. Digestive issues like constipation and gas. They can cause allergies in some people. People suffering from diabetes should consult their dietician and physician before consuming lotus seed. It tends to bring down the blood sugar levels. Makhana or Euryale Foxnut, lotus seeds, gorgon nuts and phool makhana.

These seeds are commonly used in a variety dishes and sweets, Kheer, raita and makhana curry. Makhana nutrition per 100g contains 356 Kcal 8.7 grams of protein, 7.9 grams of carbohydrates and 0.5 grams of fat. It is also a good source of various vitamins and minerals such as calcium magnesium, phosphorus and zine. Makhana has replace of popcorn. It is crunchy, tasty, simple to prepare inexpensive and healthy They are also known as lotus. They are available in a variety of qualities in the market. The size and colour the lava are very important thing. Some of the advantages of makhana include antioxidant properties and low-calorie content. It is also popularly consumed as a fasting food during Navratri in India. Makhana boasts a remarkable nutrition profile. Seed contains a low-calorie and a high fibre content, making it an ideal choice to manage weight. The rich potassium content in makhana benefits the blood pressure levels. They manages blood sugar levels because they have low glycemic index. You can eat them any time as healthy snack or even use them in drinks and dished. These seeds contain an anti-ageing, enzymes that help maintain, a youthful glow. It is a delicious food. As per the nutritionist, eating too much Makhana can cause and allergic reaction in some people. It can lead to complications such as flu, diarrhea and fever. Excessive consumption may cause constipation, bloating, or flatulence.

Pregnant women and lactating mothers should be cautious when consuming fox seeds. Over consumption may produce excessive heat in the body, which can be hazardous to their health. Makhana seeds processed with salt and spices may increase a person's risk of high blood pressure at poor quality of contaminated makhana tend to put your health at high risk. In whole, now you know that makhana's benefits and side effects. It is a versatile food made from the seeds, of a lotus seeds can create allergic reactions. It is a complete snacks, full of essential nutrients. Add around 130 gms of makhana of fox seeds to your meal, to support your dietary protein intake. You can also add them with salt, black pepper of chat masala and eat them as a snack.

Supports in cardiovascular well-ness improves blood sugar and insulin levels. Contributes to better kidney health overcomes infertility. Helps patients suffering from arthritis and Osteoporasis. So makhana eat also daily because of its high nutritional value. But excessive intake must be avoided. The nutritional value of makhana are - Fibre- 14.5 gms and preteens - 9.7.

Excessive consumption of lotus seeds may cause weight gain, nutrient imbalance, blood sugar, spikes, constipation and bad digestion. Traditionally, in Ayurveda, these seeds are known to have medicinal value or properties and are effective against bile disorder (Pitta). It is an annual aquatic herb with gigantic floating leaves and rhizomatous stem, flourishing with bounty in the muddy wetlands. It is ideally of cultivated in stagnant perennial water bodies where a water depth 1-1.5 meter is maintained throughout the year. It grows well humid and sub-humid conditions and at an air. Temperature of 20°C -35°C and humidity of 50-90% and annual rainfall 100-250 cm. Fruits become fully matured within 35-40 days of flowering Fox fox nuts are highly popular in these countries- India, China, Japan and Thailand. India's exports of fox nuts grew at 11.2% annually from 2013 to 2017 and 2017-18. Thus conclusion of makhana is primarily marketed as a pop with very limited value addition.

CONCLUSION

Nutritious organized food the potential Makhana production still remains under exploited and by at large remains till date. Lack of technological interventions absence of organized marketing, structure, inefficient supply chain and most importantly weak Socio-economic condition of Makhana. The total area under Makhana cultivation in India is estimated to be 15000 Ha. It is yields 1,20,000 MT of makhana seeds. Which after processing yields 40000 MT Makhana pop. The estimated value of the production at farmers end is Rs 250 crore and generates revenue of rupees 550 crore at traders level. Makhana is rich in antioxidants and contains several amino acids that may have entire ageing properties such as helping promotes skin, elasticity and skin health.

REFERENCES

1. Jana BR, Shrivastava A, Idris M, (2019). New Makhana (*Eurgale ferox* Salisb.) processed products for health benefits. J food process Technol 10 (789).
2. Singh AV, Singh PK, (2019). Nutritional, Ethnotherapvtics and Socio-Economic Relevance of Evryale Ferox Salisb. By Ayam Victor Singh and P.K. Singh. Life Science- Leaflets 1:22: 1104-1110
3. Jha V, Shalini R, Kumari A, Jha P Sah NK. (2019). Aqua cultural Nutritional and Therapeutic Biology of Delicious seeds of *Euryale ferox* salisb: A Minireview Current Pharmaceutical Biotechnology, 19(7): 545-555.

21

The Role of Plant Breeding in Ensuring Global Food Security: Challenges and Opportunities for Developing Nations

Nusrat Ara and Shahnaz Jamil

University Dept. of Botany, L.N. Mithila University, Darbhanga-846004, Bihar

INTRODUCTION

In an era marked by growing populations, diminishing natural resources, and climate change, the global food supply faces unprecedented challenges. At the forefront of the solution to these challenges is plant breeding, a critical tool for developing new varieties of crops that can meet the rising demand for food while also addressing environmental and socio-economic concerns. The future of food security, especially in developing nations, hinges on the advancement and widespread adoption of innovative plant breeding techniques. This article explores the significance of plant breeding in achieving food security, the challenges faced by developing countries, and the opportunities for progress.

Food is fundamental to life and survival, and its availability depends largely on the agricultural sector. As the global population is expected to reach nearly 10 billion by 2050, the demand for food will surge by 70%, putting immense pressure on agricultural systems to increase productivity while minimizing environmental degradation. Plant breeding, the science of modifying plant traits to achieve desired characteristics, plays a pivotal role in this effort. By developing crop varieties that are more productive, resilient, and adaptable, plant breeding is an essential part of the strategy to feed the world.

The focus of modern plant breeding is not only on increasing yields but also on enhancing the nutritional value of crops and improving their ability to withstand environmental stresses such as drought, salinity, pests, and diseases. This becomes particularly important as climate change intensifies, leading to unpredictable weather patterns and extreme conditions that threaten food production. With plant breeding, crops can be tailored to thrive under such adverse circumstances, ensuring a more stable and secure food supply.

In addition to environmental factors, plant breeding has emerged as a key player in the transition towards more sustainable food systems. As consumers increasingly demand plant-based and eco-friendly foods, the food industry must rely on crop varieties that are not only high-yielding but also environmentally sustainable. By promoting the development of crops with lower water and nutrient requirements, plant breeding helps reduce the ecological footprint of agriculture, ensuring that future generations can access sufficient and nutritious food without depleting the planet's resources.

While plant breeding holds enormous potential for enhancing food security, its benefits are unevenly distributed across the globe. Developing nations, which are often the most vulnerable to food insecurity, face significant barriers to implementing advanced plant breeding programs. These barriers stem from several factors, including poor infrastructure, lack of funding, inadequate research and development (R&D), and limited access to cutting-edge technologies.

In many developing countries, the agricultural sector is characterized by outdated practices and limited resources, making it difficult for farmers to adopt innovative breeding technologies. Furthermore, these nations often lack the institutional capacity to conduct research and apply the latest scientific advancements in plant breeding. This results in a reliance on traditional crop varieties that may not be well-suited to changing environmental conditions or the growing needs of the population.

Another major challenge in developing nations is the lack of investment in agricultural R&D. In wealthier countries, plant breeding research is often supported by substantial government funding and private sector investment, leading to rapid advancements in the field. However, in many developing countries, funding for agricultural research is scarce, and research institutions often struggle to keep pace with global developments. This underfunding limits the ability of local breeders to create crop varieties that can address the unique challenges faced by their regions.

The gap between developed and developing nations in plant breeding capacity also has socio-economic implications. In developing nations, where a large percentage of the population relies on agriculture for their livelihoods, the failure to adopt modern breeding practices can perpetuate poverty and food insecurity. This, in turn, leads to a cycle of dependence on food aid and imports, further undermining the self-sufficiency of these nations.

OPPORTUNITIES FOR INNOVATION AND PROGRESS

Despite these challenges, there are significant opportunities for developing nations to enhance their plant breeding capabilities and improve food security. Technological advancements in plant breeding, such as genome editing and molecular markers, offer the potential to accelerate the development of new crop varieties that are tailored to the specific needs of different regions. These technologies allow breeders to introduce beneficial traits, such as drought tolerance or disease resistance, more quickly and precisely than traditional breeding methods.

One of the most promising developments in plant breeding is the use of genome editing technologies like CRISPR-Cas9, which enable breeders to make targeted changes to a plant's DNA to enhance its traits. This technology has the potential to revolutionize crop improvement, particularly in developing countries where farmers face challenging growing conditions. By incorporating traits that enhance resilience to climate change, genome editing can help crops adapt to new environmental realities and provide more consistent yields, even in the face of extreme weather events.

In addition to technological innovations, there is a growing recognition of the need for public-private partnerships to advance plant breeding efforts in developing nations. Governments, international organizations, research institutions, and private companies must collaborate to promote the exchange of knowledge, resources, and technologies. Such partnerships can help bridge the gap between developed and developing countries by providing the necessary funding and expertise to establish robust breeding programs.

Furthermore, international organizations like the Food and Agriculture Organization (FAO) and the Consultative Group on International Agricultural Research (CGIAR) have a critical role to play in supporting plant breeding initiatives in developing nations. These organizations can provide technical assistance, funding, and capacity-building programs that enable local breeders to overcome the challenges they face. They can also facilitate the transfer of knowledge and technology from more advanced countries, helping to level the playing field.

Investing in education and training for plant breeders is another key opportunity for developing nations. By building a skilled workforce of plant scientists, these countries can enhance their ability to innovate and produce new crop varieties that are suited to their local conditions. This will not only improve food security but also contribute to economic growth and poverty reduction by increasing agricultural productivity and resilience.

CONCLUSION

The role of plant breeding in ensuring global food security cannot be overstated. As the world grapples with the dual challenges of feeding a growing population and adapting to climate change, plant breeding offers a pathway to a more sustainable and resilient food system. While developing nations face significant obstacles in harnessing the full potential of plant breeding, the opportunities for progress are immense. Through investment in research, technological innovation, public-private partnerships, and capacity-building, these nations can strengthen their agricultural sectors and improve food security for millions of people. The future of global food security depends on our ability to develop crops that can thrive in an increasingly unpredictable world. By advancing plant breeding efforts and addressing the unique challenges faced by developing nations, we can ensure that everyone has access to safe, nutritious, and sustainable food – now and in the future.

REFERENCES

1. The State of Food Security and Nutrition in the World 2021, Food and Agriculture Organization of the United Nations Rome, 2: 7-50.
2. Keith O. Fuglie, Ruben G. Echeverria (2024). The economic impact of CGIAR-related crop technologies on agricultural productivity in developing countries, 1961–2020; World Development 176: 106523.
3. Sashi Lamichhane, Sapana Thapa. (2022). Advances from Conventional to Modern Plant Breeding Methodologies; Plant Breed. Biotech. 10 (1):1-100.
4. Kunling Chen, Yanpeng Wang, Rui Zhang, Huawei Zhang, and Caixia Gao (2019). CRISPR/Cas Genome Editing and Precision Plant Breeding in Agriculture. 70:667-697.
5. Alina, Hagiu; Marinela, Bărbulescu. (2024). World Bank Activity in the Field of Agriculture; Agricultural Management. Management Agricol, 26 (2): 233.
6. Understanding heat and drought stress adaptation mechanisms in maize on the molecular level. 2021. 49:521–527.

22

Future Prospects of Hydrophytes in Health Security

Mustafa Kamal Ansari[1] and Nikhath Parveen[2*]

[1]*Department of Botany, Millat College, Darbhanga-846004, Bihar.*
[2]*Department of Biotechnology, L.N. Mithila University, Darbhanga-846004, Bihar.*

INTRODUCTION

Wetland plants have been used as food supplements and medicines by the people surrounded by wetlands for a long time. Protein, Carbohydrate, minerals, iron, phenol, ascorbic acid, amino acids, sodium, potassium, fibre, calcium, vitamins, magnesium, alkaloids, polyphenols, terpenoids, steroids, glycosides, flavonoids, phenolic acid, fatty acids etc are found in hydrophytes *Caesulia axillaris* (Pink node flower), *Colocasia esculentav* (Chembu, Taro), *Euryale ferox Salisb.*(Makhana), *Ipomoea aquatica* (Karmisaag or water Spinach), *Nelumbo nucifera* (Kamal), *Ottelia alismoides* (Kauathutthi or Ducklettuce) etc; which indicates that these hydrophytes are natural resources of nutrients. In addition of some hydrophytes *Alternanthera philoxeroide* (Kabo-napi or Jal daraga), *Centipeda minima* (Bladder Cherry or Karauda), *Sphaeranthus indicus* (Mundi or Gorakhmundi), *Cyathocline purpurea* (Gangotra), *Epaltes divaricate* (Heen mudamahana, Sevendara) are found to have biochemical activities like anti-bacterial, anthelmintic, anti-HIV, anti-allergy, antibacterial, anti-allergy, antibacterial, anti-malaria, anti-allergy, anti-cancer, diaphoretic, diuretic, stimulating expectorant, hepatoprotective, antioxidant, antimicrobial, anti-diabetic, anti-ageing, anti-hyperlipidemic and are being used to cure several diseases intestinal worms; abortifacient and applied in leucoderma, cut and injuries, diabetes, boils/burns, jaundice, sore throat, painful urination, menstrual complaints, liver disorder, stomach ache, renal diseases, spermatorrhoea, anaemia, heart disease, malaria, skin diseases etc. these activities and utilization of wetland plants enhances future prospects of wetland plants in nutrition and medicine. On taking a survey of wetlands in the Darbhanga district of Bihar, hydrophytes *Euryale ferox Salisb., Ipomoea aquatica, Nelumbo nucifera, Trapa natans or Trapa bispinosa* (Singhara) and *Ottelia alismoides* are found extensively and these macrophytes are found to have nutritional and medicinal importance and also food suppliers to local people.

MATERIALS AND METHODS

Raw materials and equipment

Fresh and healthy fruits/products of the selected plants were collected from different regions of Darbhanga district of Bihar for preliminary biochemical tests, conducted in the botany lab of Millat College Darbhanga for all necessary equipment and required chemicals were arranged at my own cost.

Survey

This work was carried out for the Darbhanga district of Bihar therefore, I visited some places (Mubby, Bela Chaur, Kusheshwersthan, Benipur, Biraul, Mahapara Chaur etc) of Darbhanga where the plants are being grown on large scale and also visited to some local markets of Darbhanga where the wetland plant's products are selling.

Methodology

A comprehensive analysis and investigation process has been adopted for this work. Some phytochemical tests (Table) were carried out for the authenticity of the work and a details study were carried out through previous and current research on the selected hydrophytes.

RESULTS AND DISCUSSIONS

Makhana (*Euryale ferox*)

Makhana is a hydrophytic/wetland plant widely cultivated in several region of the wetlands of Bihar, India, and popularly, grown in the Darbhanga district of Bihar. The fruits of Makhana are good content of bio-chemical compounds namely carbohydrates, dietary fiber, and several essential minerals like magnesium (Mg), potassium(K), and phosphorus(P). Ghosh *et al.* (2018) illustrated that Makhana's carbohydrate content makes it an excellent source of energy, and the low content of glycemic in Makhana supports its use in managing to control blood sugar levels. Sharma *et al.* (2019), described that due to the carbohydrate content of Makhana, equipped with the highly antioxidant properties, and fiber content with multi-minerals & multi-vitamins, and several biochemical activities(anti-biotic, anti-microbial, anti-diabetic, anticancer, cardio-protective etc.(Saima Tehseen et al., 2020; Das, S., Der, P., Raychaudhuri, U. et al., 2006) it has been placed as a functional food suppliers with potential health securities.

Kumar *et al.* (2020) investigated the therapeutic behaviors of fruits of Makhana and observed an effective role of Makhana in cardiovascular health issues due to its low content of fat and high content of fiber. Further, he observed that the applications of Makhana in weight management and, in metabolic syndrome as well, could be an effective solution. The anti-inflammatory properties in the fruits of Makhana were linked to its bioactive compounds namely flavonoids and polyphenols, which signs as an addition to its therapeutic potential.

The utilization index of Makhana increasing fast because of the multi-nutritional benefits in Makhana, also having low calories index, as well as low bad fat index, and the higher the index of mineral in Makhana. Including the products of Makhana seeds (popped Makhana or powder of Makhana seeds mixed with some other food items in dry or jelly form) into the daily diet plan, could be an effective supplement for several diseases caused by micro-nutrient, vitamins & minerals deficiencies. Also, the fruits of fox nuts has HBAC content which has been tested as a chemical compound to release insulin internally in human body from the pancreatic(β-cells) tendency towards improve in sugar level glycemic control(Joseph, A. et al. 2023). The least value of glycemic index in the fruits of Makhana is a positive signal for its utilization against several chronic-diseases. The lower glycemic indexed foods are obviously recommended by FSSI(Liyakat, M. et. al. 2022). Makhana nuts having a large list in ancient medicinal literature of China, Nepal and India.

Makhana fruits contains multivitamins namely proteins, Polyphenol (Resveratrol), fibres, carbohydrates, Steroid compounds like β-sito-sterol , Dauco-sterol , Fuco-sterol, Magnesium, Zinc, Potassium, Phosphorous together with Essential oil namely Furfural, Penta-noic acid, furan-carboxaldehyde, Iso-creosol , 2-methylphenol , (1, 2, 3-trimethoxybenzene), Palmitic acid, Pentamethyl benzene, 4-ethylguaiacol and several Amino acids. (Jiang, J. et. al. 2023) but very low content of sodium and cholesterol. (Devi KR., 2022), Kumar, M et al., 2021).

Singhara (*Trapa bispinosa*)

Singhara, also called 'bull nuts' is found in watery region many district of Bihar including Darbhanga. It has been utilised in food materials locally in Darbhanga however its bio-chemical analysis enhances its nutritional and medicinal utilization. Presence of multi-vitamins and multi-minerals in Singhara namely iron(Fe), copper(Cu), carbohydrates, calcium(Ca), phosphorous(P), manganese(Mn), sodium(Na), potassium(K), magnesium(Mg) (Bharthi, V. et. al. 2015), multi-vitamins namely thiamine, riboflavin, vitamin E,C & A, Vitamin D-amylase, nicotinic acid Vitamins B-1, B-2, B-5 and B-6 and phosphorylases, citric acid, lipids and crude fibres. (Vageshwari and Pandey V. N. 2019; Singh, G. D. et. al. 2010; Basu, S. et al. 2023). A number of phytonutrients, also including the biochemicals catechins and epicatechins signs that something antioxidants are available in it.

Singhara fruits has been tested positive for several bio-chemical activities, regarding its medicinal utilization for human-being. Bio-chemical activities such as anti-oxidant, anti-diabetic, Analgesic, Anti-microbial & Anti-bacterial activities, Antifatigue activity, Anti-ulcer & anti-cancer activity, Anti-depressant activity, Cardio & Neuroprotective activity, Hepato-protective activity & Anti-inflammatory. [Rani, B. et. al; 2016.; Corovic, R. C. et al. 2020.; Joshi, S. et al. 2021.; Bharthi, V. et. al. 2015.; Basu, S. et al. 2023.; Rajput J. D. et al. 2023.; Singh, G. D. et. al. 2010].

Singhara plant products are very useful against several diseases namely polyuria, hair diseases, diarrhoea, dysuria, weakness, skin problems, fever, sexually-deficiency diseases, sore throat, body aches etc in Ayurvedic as well as Homeopathy and are also utilized for nutritional benefits (Basu, S. et al. 2023). A local survey reported that its bread is being served to pregnant ladies during delivery showing its utilization against blooding.

Kamal (*Nelumbo nucifera* Gaertn.)

Indian lotus, a wetland plant, known by KAMAL as the 'Sacred flower' that is because of its significance and important role in several religious ceremonies. It is used as an ornament as well as offerings in temples and secret places of Hindus. It is bestowed as the 'National' Flower of India. Lotus plants can be grown in tanks, ponds, muds, marshes etc, starts in ending of winter, and last up to the whole spring season, planted either by rhizomes or by seeds(Lakshman Chandra De. 2020). Lotus plants is counted in vegetables and flowers have been used as a medicinal plant for many years due to their multi-health protective bio-chemical ingredients (Lakshman Chandra De. 2020). Lotus plants were first introduced in the watery area of Europe (water-lily family member) in the early 18th century, however, this plant is 'native' to India, now spread in many countries of Asia, Persia and Egypt. Seeds, flowers, and rhizome of the lotus plant are taken as food items. The lotus seeds are eaten as raw or roasted or ground or boiled form. The roasted form of Kamal can also be taken instead of coffee as it contains saponins and phenolics ingredients(Moro, C. F. et. al. 2012). Lotus seeds as well as plants are highly nutritious content, having good quantity of proteins, multivitamins, lipids, carbohydrates, and multi-minerals(Bangar, S. P. et. al. 2022). The lotus plants (rhizomes, seeds & stems) have biochemical ingredients proteins, carbohydrates, calcium(Ca), iron(Fe), zinc(Zn), fat, crude fibre; multi-mineral sodium(Na), potassium(K), phosphorus(P), magnesium(Mg), copper(Cu), manganese(Mn), selenium lipids, poly-unsaturated fatty acids, saturated fatty-acids, oleic & cotyledon and tannin content(Moro, C. F. et. al. 2012) and multi vitamins namely vitamin-B(also B1, B2, B6), Vitamin-C & E and also alkaloids content (Bangar, S. P. et. al. 2022).

Lotus plant's part and seeds were found to have biochemical activities anti-cancer, gastro-intestinal regulation activity, immune-modulatory activity, anti-inflammatory, antioxidant, antifibrosis, Anti-ischaemic activity, Diuretic effect, Hepatoprotective. (Hajela, M. S. et. al. 2018; Bangar, S. P. et. al. 2022). Therefore the Kamal plants could be utilized as effective drugs against cancer, skin diseases, diabetes, tissue inflammation, bleeding problems & cardiac diseases etc[Hajela, M. S. et. al. 2018; Bangar, S. P. et. al. 2022; Moro, C. F. et. al. 2012].

Table-1

Si. No.	Biochemical Test	Content	Makhana	Singhara	Kamal
1.	Benedict's test	Carbohydrates	+	+	+
2.	Biuret's test	Protein	+	+	+
3.	Borntrager's test	Glycosides	+	+	-
4.	Acrolein test	Fats	+	+	+
5.	Benedict's test	Reducing Sugar	+	+	+
6.	Millon's test	Amino acids	+	+	+
7.	Mayer's test	Alkaloids	+	-	+
8.	Shinoda test	Flavonoids	+	+	+
9.	Iodine test	Starch	+	+	+
10.	Ferric chloride test	Phenols	+	+	+

CONCLUSION

The present study is an observation for the importance of the hydrophytes Makhana, Singhara and Kamal of Darbhanga district of Bihar in health securities like 'medicinal, nutritional and food items. Without hesitation, these plant fruits could be treated as 'dry fruits' as they have rich content of multivitamins and nutrients like proteins, carbohydrates, amino acids and many more. They showed several biochemical activities that enhances their drug like recognition against several chronic and normal diseases. An effective steps towards the industrialization of several product made from Makhana, Singhara and Kamal, alone or mixed with some another dry fruits, edible items could be a good idea. The products like biscuits, puffed Makhana/lotus seeds like chips packs, powdered immuno-booster, drinks, street food items/Chat like Chinese foods, Pests etc can be made from Makhana, Singhara and Kamal. Medicine as well as food industry could be grown through Makhana, Singhara and Kamal in Darbhanga district of Bihar.

REFERENCES

1. Adkar P, Amita Dongare, Shirishkumar Ambavade, and V. H. Bhaskar (2014). *Trapa bispinosa* Roxb.: A Review on Nutritional and Pharmacological Aspects. Advances in Pharmacological Sciences. http://dx.doi.org/10.1155/2014/959830

2. Ahmed D, Vikas Kumar, Amita Verma, Girija Shankar Shukla and Manju Sharma (2015). Antidiabetic, antioxidant, antihyperlipidemic effect of extract of *Euryale ferox* Salisb. with enhanced histopathology of the pancreas, liver and kidney in streptozotocin-induced diabetic rats. Springer Plus 4:315. DOI 10.1186/s40064-015-1059-7

3. Alfasane, M. A., Moniruzzaman Mhondker and M. Mahbubar Rahman (2011). Biochemical composition of the fruits of water chestnut (*Trapa bispinosa* Roxb.). Dhaka University Journal of Biological Sciences. 20(1): 95-98.

4. Alfasane M. A., Moniruzzaman Khondker, Z. N. Tahmida Begum, Laila Arjumand Banu, Md. Mahbubar Rahman and Umma Fatema Shahjadee (2008). Fruit production and biochemical aspects of seeds of *Euryale ferox* Salisb. Under ex-situ conditions. Bangladesh Journal of Botany. 37(2): 179-181.

5. Atal R., Shridhar Patil, R. K. Sohane, Ram Datt, Suborna Roy Choudhary and S. N. Singh (2019). Economics of Makhana Production and Socioeconomic Profile of Makhana Growers in Bihar. Current Journal of Applied Science and Technology. 38(6):1-5

6. Bangar, S. P., Kyle Dunno, Manoj Kumar, Hussein Mostafa and Sajid Maqsood (2022). A comprehensive review on lotus seeds (*Nelumbo nucifera* Gaertn.): Nutritional composition, health-related bioactive properties, and industrial applications. Journal of Functional Foods. 104937(89). https://doi.org/10.1016/j.jff.2022.104937

7. Basu S, Hemamalini A. and Janardhanam (2023). Nutritional and Health Benefits of Singhara: A Review Article. International journal of health and allied sciences. 12: 66-70.

8. Bharthi V, Kavya B, Shantha TR, Prathapa Reddy M, Kavya N, Rama Rao V, Kalpesh Kumar B, Ishnawa, Venkateshwarlu G (2015). Pharmacognostical evaluation and phytochemical studies on Ayurvedic nutritional fruits of *Trapa natans* L. International Journal of Herbal Medicine. 3(5):13-19.

9. Bishayee A., Patel P. A., Sharma P, Thoutireddi S, Das N (2022). Lotus (*Nelumbo nucifera* Gaertn.) and its Bioactive Phytocompounds: A Tribute to Cancer Prevention and Intervention. Cancers (Basel). 14 (3): 529. DOI: 10.3390/cancers14030529.

10. Corovic R. C., Jovana Bradic, Marina Tomovic, Vera Dabanovic, Vladimir Jakovljevic, Gala Zarkovic and Zeljka Rogac (2020). Chemical composition and biological activity of *Trapa natans* L. DOI: 10.2478/sjecr-2021-0032.

11. Devi KR (2022). A Comprehensive Review on Nutraceutical Properties of *Euryale ferox* Salisb. Journal of Plant Science Research. 9 (2): 222.

12. Hajela, M. S., Dr. Rachana Srivastava, Ms, Jasnoor Kaur (2018). A Review on Health Benefits of Lotus Seeds (*Nelumbo nucifera*). Journal of Applied Science and Computations. 5 (4): 92-96.

13. Jiang J, Haiyan Ou, Ruiye Chen, Huiyun Lu, Longjian Zhou, and Zhiyou Yang (2023). The Ethnopharmacological, Phytochemical, and Pharmacological Review of *Euryale ferox* Salisb.: A Chinese Medicine Food Homology. Molecules. 28: 4399.

14. Jiang J, Haiyan Ou, Ruiye Chen, Huiyun Lu, Longjian Zhou, Zhiyou Yang (2023). The Ethnopharmacological, Phytochemical and Pharmacological Review of *Euryale ferox*, A Medicine Food Homology Species. dOI:10.20944/preprints 202305.0340.v1.

15. Joseph A. and Gowri Ramesh (2023). Nutrient Analysis, Phytochemical and Antioxidant Activity of a Food Product Formulated with Fox Nuts (*Euryale ferox*). Asian Journal of Biological and Life Sciences. 12 (2): 279-285.

16. Liaquat M, Imran Pasha, Muhammad Ahsin and Abdullah Salik (2022). Roasted fox nuts (*Euryale Ferox* L.) contain higher concentration of phenolics, flavonoids, minerals and antioxidants, and exhibit lower Glycemic Index (GI) in human subjects. Food Production, Processing and Nutrition. 4:1. https://doi.org/10.1186/s43014-021-00081-x

17. Lin Z., Cheng Zhang, Dingding Cao, Rebecca Njeri Damaris and Pingfang Yang (2019). The Latest Studies on Lotus (*Nelumbo nucifera*)-an Emerging Horticultural Model Plant. International Journal of Molecular Sciences. 20, 3680; doi:10.3390/ijms20153680 .

18. Mann S., Deepika Gupta, Velol Gupta and Rajinder K. Gupta (2012). Evaluation of nutritional, phytochemical and antioxidant potential of *Trapa bispinosa* roxb. Fruits. International Journal of Pharmacy and Pharmaceutical Sciences. 4 (1): 432-436.

19. Meena S. K., Deepa H. Dwivedi, Nirmal Kumar Meena, Monika Bhanwariya, Krishna Jat, Tendul Chouhan and Bhupendra Meena (2023). Proximate and Biochemical Analysis of Water Chestnut (*Trapa natans, var. bispinosa* Roxb.) kernel as affected by various inorganic and organic sources. Biological Forum-An International Journal. 15(10): 398-402.

20. Rajkumar P. and Madhiraju Rajithasri (2022). Water chestnut: Growing conditions, nutritional and phytochemical composition, novel extraction methods and health properties. The Pharma Innovation Journal. 11(7):599-612 .

21. Rajput J. D., Sury Pratap Singh (2023). Water Chestnut (*Trapa natans* L.): Functional characteristics, nutritional properties and applications in food industry: A review. The Journal of Phytopharmacology. 12(2):119-126.

22. Rani B., Deepmala Verma, Itishree Bhati, Harsukh Chharang, Raaz K Maheshwari (2016). Health Benefits of Scrumptious Water Chestnuts /Water Caltrop (*Trapa natans* L.). International Archive of Applied Sciences and Technology. September. 7 (3): 32-36.

23. Kumar R., Dr. Mohit Sanduja, Neelam Painuly (2024). A Research on Phytochemical and Antimicrobial Investigation of *Euryale Ferox* Seeds. Educational Administration: Theory and Practice. 30(6): 3476-3482.

24. Rather G. J., Hamiduddin, Mohd Ikram, Shaista Fatima, MD Naquibuddin (2018). Physicochemical Standardization of Polyherbal Powder Formulation: Safoof-e-Makhana. Pharmacognosy Journal, 10 (5): 899-906.

25. Saeed, S. M. G., Sidra Tayyaba, Syed Arsalan Ali, Sania Tayyab, Syed Asad Sayeed, Rashida Ali, Lubna Mobin & Shahina Naz (2020). Evaluation of the potential of Lotus root (*Nelumbo nucifera*) flour as a fat mimetic in biscuits with improved functional and nutritional properties. Cyta - Journal of food. 18 (1): 624–634. https://doi.org/10.1080/19476337.2020.1812727.

26. Sahu, R., J. Singh, Nandan Mehta, T. Tirkey and R. R. Saxena (2023). Role of Coconut Water, Gibberelic Acid in Vase Life of Existing Lotus (*Nelumbo nucifera* Gaertn) Genotypes in Chhattisgarh Plains. International Journal of Current Microbiology and Applied Sciences. 12 (4). https://doi.org/10.20546/ijcmas.2023.1204.025.

27. Sharma A, R. R. Mishra, Bipin Saw and Nasim Ahmad (2020). Economic Analysis of Makhana Cultivation in Darbhanga and Madhubani Districts of Bihar, India. International Journal of Current Microbiology and Applied Sciences. 9 (4) 1097-1102. https://doi.org/10.20546/ijcmas.2020.904.130

28. Singh G. D., Sukhcharn Singh, Navdeep Jindal, Amrinder S. Bawa and Dharmesh C. Saxena (2010). Physico-chemical characteristics and sensory quality of Singhara (*Trapa natans* L.): An Indian water chestnut under commercial and industrial storage conditions. African Journal of Food Science. 4(11): 693-702.

29. Sodi M., Sanjib Kumar (2019). Makhana in Nutritional, Medicinal and Socio - Cultural Uses in the Koshi - Mithila Region of Bihar. Scholars Academic Journal of Biosciences. 7 (6): 276-287.

30. Tehseen S., Fatima Sarfraz, Sidra-tul-Muntaha, Naila Ateeq, Faiza Ashfaq, Iqra Yasmin and Tayyaba Mehmood (2020). Foxnut (*Euryale ferox* Salisb.): A Health Promising Fruit. Acta Scientific Agriculture. 4 (12): 68-72.

31. Vageshwari and Pandey V. N. (2019). Phytochemical and Therapeutic Potential of Bioenergetic plant: *Trapa bispinosa* Roxb. Journal of Drug Delivery & Therapeutics. 9(2):111-114. http://dx:.doi.org/10.22270/jddt.v9i2-s.2470

23

Challenges and Opportunities in Maize (*Zea mays* L.) Production in India: A Review

Mahesh Kumar, Lucky Rani, Diksha Sultania and Nitu Kumari

University Department of Botany, L.N. Mithila University, Darbhanga-846004, Bihar

INTRODUCTION

Maize, scientifically known as *Zea mays L.*, is the third most significant cereal crop in India, ranking behind rice and wheat. The cultivation of this crop spans across 9.09 million hectares, yielding a yearly output of 24.26 million metric tonnes. The average national productivity stands at 2.56 metric tonnes per hectare, as reported by Yadav *et. al.*, in 2015. Maize cultivation takes place during the monsoon (kharif), winters (rabi), and spring seasons, with the rainy season being the main period of output. It is grown in nearly all regions in the country due to its extensive ecological tolerance. The states that produce the most maize are Andhra Pradesh (20%), Karnataka (17%), Maharashtra (11%), Bihar (9%), Tamil Nadu (8%), Madhya Pradesh (6%), Rajasthan (6%), and Uttar Pradesh (5%).

Maize is the primary feed grain in India, with around 59% of the total production being used as feed. The remaining portion is distributed among industrial use raw material (17%), food (10%), exports (10%), and miscellaneous purposes (4%) (Kumar *et. al.*, 2013). It is considered a globally important commodity that plays a crucial role in driving global agriculture due to its wide range of uses in the feed, factories, and food sectors. It is grown in 165 countries across 184 million hectares worldwide. The total production of maize is 1,016 million metric tonnes, with an average production rate of 5.52 metric tonnes per hectare (FAOSTAT, 2014).

Since 1996, this grain has surpassed rice and wheat to become the most-produced grain in the world. Its production rate is currently increasing at double the rate on an average of rice and three times the rate of wheat (Fischer *et. al.*, 2014) in Fig.1.

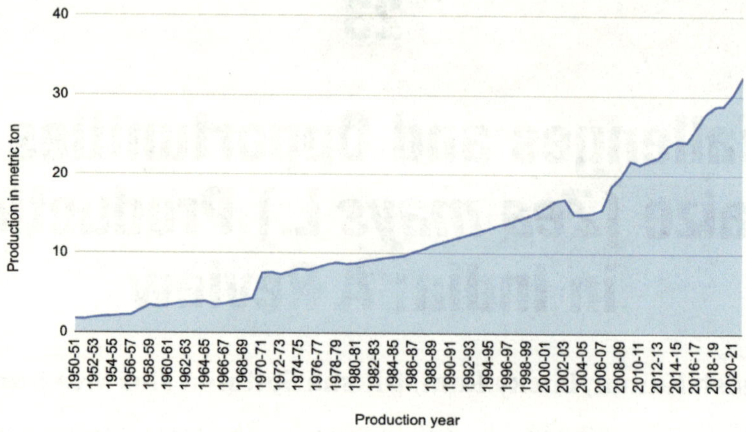

Fig. 1. Maize output trends in India from 1950-51 onwards The source of the information is the "Agricultural Statistics at a Glance 2021" report published by the Ministry of Agriculture, Government of India.

The above graph analysis examines the need, potential, and difficulties in significantly boosting Indian cultivation of maize to around 50-55 million metric tonnes by 2030. To achieve this objective, an annual growth rate (CAGR) of 7.18% is necessary. During the period from 2003-04 to 2023-24, maize output in India showed a Compound Annual Growth Rate (CAGR) of 4.98%.

Hence, the durable technological interventions and a significant demand will be essential in order to accomplish the goal of doubling of maize production in India.

MATERIALS AND METHODS

The analysis is based on secondary data collected from agricultural reports, scientific studies, and expert consultations. Key sources include the "Agricultural Statistics at a Glance" by the Ministry of Agriculture, FAOSTAT (2014). The study evaluates maize productivity trends, technological advancements, and policy frameworks influencing maize cultivation in India. The methodology involves:

1. **Data Collection**: Historical and current data on maize production, demand, and area of land.
2. **Trend Analysis**: Evaluation of production trends from 1950 to 2023 to project future growth rates.
3. **Demand Assessment**: Analysis of Sector-Specific demand, including food, feed, industrial, and export requirements.
4. **Policy Review**: Examination of existing policies and their impact on maize production.

5. **Technological Assessment**: Review of advanced technologies like hybrid seeds, conservation agriculture, and stress-resilient breeding techniques.

Factors Influencing Future Demand for Maize

A. The need for feed: Maize is a very suitable feed for chicks and is anticipated to continue being a fundamental component in industry. Presently, a total of around 15 million metric tonnes of maize grain is utilized as animal feed, with a specific allocation of about 12.5 million metric tonnes solely for chicken feed. According to Anonymous (2020), the Indian chicken feed industry is forecasted to see a yearly growth rate of 7-8%. Additionally, the total feed market is likely to rise at a compound annual growth rate (CAGR) of 8%.

B. The demand for exports: In recent times, India has become a major exporter of maize grain. Amounted to 4.8 million metric tonnes in the 2012-13 period. The majority of India's maize exports, accounting for 85-90% of the total, are directed towards South east Asian nations such as Indonesia, Vietnam, Malaysia, and Taiwan. The remaining 10-15% is shipped to Middle East countries like the United Arab Emirates, as well as South Asian countries including Bangladesh, Nepal, and Bhutan. Asian countries collectively represent over 50% of the worldwide maize imports, amounting to approximately 100 million metric tonnes. India gains, from both cost and transportation advantages in this extensive global maize market place (Kumar *et. al.*, 2013). India is also a prominent exporter of maize in Southeast Asia, namely in Vietnam and Indonesia, among the 11 Asian countries that buy over a million metric tonnes of maize.

India has the potential to export approximately ten million metric tonnes of maize by the year 2030, as long as the local demand is well supplied in several industries, the total demand for maize is anticipated to be strong in the next ten years. It is expected to have a demand of 50 million metric tonnes of maize by India is 2030. Out of this, 32 million metric tonnes would be needed for animal feed, 15 million metric tonnes for industrial, two million metric tonnes for food consumption, and a million metric tonnes for seed and various other purposes. By 2030, India would need to produce over 50-55 million metric tonnes of maize in order to meet both local and export requirements.

C. Food demand: According to the NSSO- 2014 report, the average per capita annual consumption of maize in rural areas in 2021-22 was 2.36 kg per person, whereas in urban areas, it was 0.268 kg per person. According to estimates derived from the census taken in 2011 data, the quantity of maize used as food for humans directly reached 4.36 million metric tonnes in the period of 2021-22. The historical consumption surveys done by the NSSO indicate a substantial decrease in rural maize consumption.

During the year 2011-12, the consumption of maize as a food source was significantly higher in states like Himachal Pradesh, Gujarat, Jammu & Kashmir, Madhya Pradesh, and Rajasthan. However, due to the rising interest among city dwellers in specialty corn products like sweet corn, baby corn, popcorn, and quality protein maize (QPM), as well as the growing popularity of multi-grain flour, it is expected that the demand for maize as a food source could reach 50 million metric tonnes by 2030.

D. Industrial demand: In India, the starch industry is now in its initial phase, with the average amount of starch consumed per person per year being only 1.3 kg. This is significantly lower compared to the United States, where the average consumption is 64.5 kg, and other Asian Nations like China and Indonesia, where it exceeds 10 kg. Nevertheless, it is anticipated that this number will increase in the future as starch is utilized in various sectors such as dietary supplements, paper, pharmaceuticals, textiles, biofuels, and in others. Due to the rising need for starch products in different industries, Indian starch industry is expected to growth rate of around 15% in the upcoming years (Anonymous, 2022). It is expected that the requirement for maize for industrial applications will increase from 13.25 million metric tonnes to 25 million metric tonnes by 2030.

Challenges in Increasing Maize Production

Drought is recognized as the primary limitation in rainfed areas, which make about two-thirds of the cultivation of maize in India. The difficulty is anticipated to escalate in the upcoming decades as a result of climate change, frequently resulting in inadequate and/or irregular precipitation during the agricultural season, combined with swings in temperature (Bernstein et. al., 2007). India is currently witnessing a rise in the occurrence and intensity of extreme weather conditions, such as floods, heatwaves, and powerful gusts of wind (Prasanna et. al., 2014, Diksha, 2022).

The impact of climate change not only increases the occurrence and strength of non-living environmental pressures. Maize in India, is currently vulnerable to various diseases and pests, including post-flowering stalk rots (PFSR), turcicum leaf blight (TLB), banded leaf and sheath blight (BLSB), downy mildews, ear rots, stem borers, and weevils. Additionally, there is a significant occurrence of mycotoxins (Lucky *et. al.*, 2023) in maize grain harvested in specific regions.

Maize is highly susceptible to heat stress, specifically during its reproductive phase. Climate forecasts indicate that increased temperatures, particularly in places susceptible to drought and reliant on rainfall for agriculture, are expected to cause substantial reductions in crop yields (Cairns *et. al.*, 2013). Spring maize, a crucial choice for enhancing and broadening agricultural practices in India, is particularly susceptible to extreme heat stress during the blooming and early grain-filling phases, especially in the upper and intermediate Indo-Gangetic plains. Therefore, it is crucial to generate enhanced maize varieties that possess inherent resistance to stress caused by high temperatures (Cairns *et. al.*, 2013; Yadav *et. al.*, 2015).

Poor soil fertility, specifically deficiency in micronutrients in degraded soils, and low nutrient use efficiency are significant factors that restrict maize yields and productivity stability in both high-potential-low-risk as well as low-potential-high-risk environments, in addition to abiotic and biotic stresses.

The limited acceptance of advanced production technologies continues to be a major obstacle to improving maize output in India. Presently, a mere 65% of the maize area of land is cultivated with high-yielding varieties. Inadequate agronomic management techniques result in the underutilization of the potential for production of these cultivars. The presence of weeds during the period known as kharif poses the most major obstacle in maize-based agricultural systems. Moreover, the Indian corn sector faces an extra hurdle caused by insufficient drying facilities, leading to post-harvest losses.

Strategies and Agricultural Innovations to Enhance Maize Yield

In the future, it will be crucial to implement durable interventions in order to achieve a significant rise in production of maize. For getting this objective, it will be imperative to maintain a yearly growth rate of 7-8%. The main emphasis should be on enhancing production rather than growing the area of land. Aiming for a productivity of around 7-8 tonnes per hectare is essential for achieving a two fold increase in production without the need of extra land. Nevertheless, the high demand is expected to result in a moderate increase in the cultivation of maize. It is estimated that by the year 2030, the total area of land dedicated to growing maize would expand to 18 million hectares. Kumar *et. al.,* (2013) forecasted a substantial increase in the cultivation of maize in the states of Karnataka, Maharashtra, Odisha, Andhra Pradesh, and Tamil Nadu. The ecological conditions for growing maize in these states are highly conducive, as evidenced by a substantial increase in maize area of land ranging from 141% to 445% during the previous two decades in these locations. Over the previous twenty years, the area has increased in both the kharif and rabi seasons, with a total of 1.8 million hectares added during the period known as kharif and 1.2 million hectares added during the rabi season. Nevertheless, the expansion of land area has shown a significantly higher rate of increase during the period known as rabi (267%) compared to the season of kharif (33%). This suggests that maize cultivation is progressively expanding in more favourable environments.

The maize productivity levels in several prominent maize-producing nations, including the United States of America (10 tonnes/ha), the Chinese mainland (6.0 tonnes/ha), Brazil (5.3 tonnes/ha), the Argentina (6.6 tonnes/ha), and neighbouring South Asian countries like the Republic of Bangladesh (6.6 tonnes/ha) and the Pakistan (4.2 tonnes/ha), indicate that there is substantial scope to enhance India's maize productivity from its current level of 2.56 tonnes/ha. In order to double maize output, it is essential to implement both technological and policy measures to reach the desired level of productivity.

Policy Interventions

By using appropriate policy initiatives strategies, production can be significantly enhanced by effectively bridging the yield gap through the use of current technologies. Implementing strategies such as the utilisation of single-cross hybrids, the authorisation of genetically engineered maize for traits essential to farmers. The promotion of scientific farming practices, the creation of mechanisms to stabilise prices, and the offering of rewards for maize-based manufacturing sectors can significantly boost maize production in India's.

Emphasis on the Use of Single-Cross Hybrids: Several single-cross variants have been introduced for cultivation, with yield potentials over 12 tonnes per hectare. Nevertheless, the current cultivation of hybrids encompasses about 25-30% of the entire maize area of land. In order to increase production, it is crucial to expand the use of single-cross hybrids that are specifically designed for specific specialized adaptations. Seed companies should focus their product offerings on certain geographic regions where each hybrid is best suited. Governmental extension agencies should actively participate in the distribution of single-cross hybrids by seed subsidy schemes and organizing on-site demonstrations on farmers' farms.

Enhancing Collaborations between the Public and Private Sectors: Both the governmental and private sectors are involved in the manufacturing and distribution of maize seeds. Seed production in the public sector involves the participation of the National Seeds Corporation, 15-State Seed Corporations, different State Departments of Food and Agriculture, and Agricultural Universities.

To enhance the manufacturing and commercialization of hybrids developed by public sector research institutions, it is necessary to enhance the collaboration between these institutions and small private seed firms through public-private partnerships (PPPs).

Dissemination of Scientific Agricultural Practices: The disparity in crop productivity between various locations that cultivate maize emphasizes the necessity for durable extension endeavors. It is essential to spread enhanced production technology, such as scientific agricultural practices, in regions optimal agricultural techniques, such as the utilization of superior seeds, efficient soil nutrient & tillage management, increased planting densities, the adoption of farm machinery, implementation of plant protection measures, and proper handling after harvest.

Genetically modified maize policy: Maize producers in other countries have extensively utilized efficient genetically modified (GM) technology to manage weeds and control stem borer for almost last 20 years. India has been conducting adaptive studies of herbicide-tolerant & insect-resistant maize cultivars for the past nine years. India requires a pragmatic and future-oriented approach towards genetically modified (GM) crops. This program would offer

Indian farmers additional choices for managing their crops, similar to what farmers in other countries have. Establishing a definitive policy for genetically modified (GM) maize, as well as GM crops in broad terms, would also facilitate continued scientific endeavors to enhance other desirable characteristics, such as droughts and disease tolerance.

Enhancing the infrastructure for handling and storing agricultural product after harvest: India is experiencing a substantial deficiency of adequate post-harvest infrastructural amenities. Approximately 2.45% of maize goes to waste at the farmers' level during various stages of the agricultural process, including harvesting, threshing, winnowing, shipping, and storage. Karnataka, the leading state in maize agriculture in India, experiences a post-harvest loss of approximately 3.02% at the farm level. Losses are incurred at various stages: during storage (0.68 kg/quintal), during drayage (0.49 kg/quintal), during transportation (0.44 kg/quintal), during threshing (0.34 kg/quintal), during packaging (0.15 kg/quintal), and cleaning (0.10 kg/quintal). In order to tackle these issues, it is imperative to build a network of community-based dryers with the aim of enhancing the overall quality of the final grain.

Market stabilization systems : It is crucial to provide maize farmers with profitable prices in order to encourage them to embrace modern growing methods and boost the nation's maize production. Maize has predominantly flourished based on its inherent capabilities and the forces of supply and demand, unlike rice and wheat. While the minimum support price (MSP) is declared on a yearly basis, the public procurement of maize is limited due to its exclusion from the system of public distribution (PDS). The prevailing market price of maize has consistently exceeded the officially declared Minimum Support Price (MSP) owing to robust demand. An important factor in the maize market in India is its significant presence in the export market, with an annual export of 3-4 million metric tonnes of maize. The export volume is contingent upon the current international market prices. Countries that produce and export large amounts of maize are embracing technical advancements and implementing measures to increase productivity while lowering costs. Unless remedial steps are implemented, the competitiveness of Indian maize may be further diminished due to the devaluation of currencies in many countries, reduced yields of Indian maize, and a strict regulatory environment. In 2022-23, the volume of maize exports from India experienced a notable decline compared to prior years. As a result, it became necessary to establish systems that would provide compensation to farmers during periods of price crashes, with the backing of government-backed price stabilization funds.

Maize-based processing industries are being offered incentives : The maize processing sector in India is very underdeveloped compared to countries such as the U.S. and China. There is a substantial opportunity to enhance the utilization of maize in industrial uses. Maize starch serves as an adhesive in the textile business, a thickening agent in the food industry, a means to enhance paper strength in the paper industry, a filler in the pharmaceutical industry, and a raw material for the manufacturing of glucose, dextrose, ethanol, and various other products. In India, maize wet grinding is the main method used for starch processing. Around 40% of all enterprises involved in the manufacturing of starch are new participants, necessitating increased incentives to develop a strong maize starch industry. Expansion in the starch sector would result in a rise in the need for maize and, consequently, greater profits for farmers.

Technological Interventions in Maize Cultivation

Germplasm diversification and heterotic classification of germplasm: The lack of genetic diversity in crops, like maize, makes them more susceptible to diseases, pests, and environmental pressures, ultimately leading to reduced yields. Therefore, it is crucial to consistently incorporate novel germplasm from various origins into maize breeding projects. This infusion is crucial for mitigating genetic susceptibility and optimizing performance in all characteristics, including crop production and its components. Expanding the genetic diversity of the germplasm base can provide new and unique characteristics, ultimately benefiting for farmers.

An initiative similar to the Germplasm Enhancement of Maize (GEM) project in the USA (Blanco and Gardner, 2014) would be crucial for the advancement of Indian maize breeding initiatives. Currently, there is a scarcity of information regarding heterotic patterns in Indian maize germplasm, with the exception of a few ground breaking studies carried out at the Punjab Agricultural University (Dhillon *et. al.*, 1997). Categorizing maize-inbred lines into separate heterotic groupings might accelerate the utilization of increased levels of heterosis. Enhancing the genetic material foundation is crucial for the ongoing advancement of superior hybrids, ultimately resulting in heightened maize yields.

Advancement of Germplasm with Enhanced Stress Tolerance: At present, there are maize hybrids that may produce 12-14 tonnes per hectare, but these high yields are usually only possible when the plants are grown in ideal conditions without any stress and with the use of a lot of resources. Maize fields in India are experiencing more frequent droughts, higher temperatures, occasional severe rainfall or flooding, and the appearance of novel illnesses and insect pests. The development of climate-resilient cultivars is crucial for the future of maize production, especially during the kharif season, and for the livelihoods of millions of small-scale farmers who depend on maize for their profits.

It is necessary to create screening sites that can handle stress and establish standardized methods that are specifically designed for the projected stress combinations in the target environments (Prasanna *et. al.*, 2014). Recent developments in plant genetic engineering, RNA interference, and targeted mutagenesis methods offer promising prospects for modifying maize germplasm to enhance its resilience against both biotic and abiotic stressors (Castiglioni *et. al.*, 2008).

Improve Maize Breeding with Innovative Tools and Approaches: It is crucial to have the capacity to efficiently and economically create superior maize hybrids that have a high potential for yield and possess the necessary adaptation characteristics, such as the ability to withstand abiotic and biotic challenges. Although conventional breeding has been effective in creating a variety of high-quality maize hybrids, recent advancements in breeding methods and tools, such as doubled haploidy (DH), mechanization of breeding processes, molecular marker-assisted reproduction, high-throughput phenotyping, and year-round nurseries, along with decision support tools, present significant opportunities for enhancing genetic improvements and breeding efficiency (Prasanna *et. al.*, 2014).

The DH method allows for the quick generation of fully homozygous maize lines, presenting great prospects for the expedited creation and introduction of superior cultivars (Prasanna 2012). Integrating DH lines with molecular markers greatly enhances genetic gains and boosts breeding efficiency. Currently, only a limited number of breeders or institutes in India are employing DH technology and molecular markers. It is crucial to promptly incorporate these technologies throughout maize breeding efforts, especially in the public sector.

Genome-wide association studies (GWAS) are a strong method for analyzing complex features and detecting advantageous genetic variations in maize. This is achieved by the use of high-throughput genotyping and precise phenotyping techniques (Prasanna *et. al.*, 2014). An organized and focused endeavor is necessary to create markers that are ready for use in breeding programs to build resistance to significant maize diseases in India. Molecular marker-assisted breeding is crucial for addressing the task of creating cultivars with certain adaptation characteristics, such as tolerance to biotic and abiotic stress (Prasanna *et. al.*, 2014).

It is now possible to use inexpensive and user-friendly equipment to study specific features in the field. In addition, it is necessary for public and commercial organizations to cooperate together in order to create "phenotyping networks" that can thoroughly and effectively analyze genetic resources and breeding materials for various desired traits, especially those related to tolerance of biotic and abiotic stresses and nutritional quality. This partnership would greatly expedite genomic-guided breeding, enhance the genetic diversity of high-quality breeding resources, generate innovative cultivars, and mitigate the impacts of climate change.

Technologies for Conservation Agriculture: Conservation Agriculture (CA) technologies offer the chance to cultivate full-duration maize hybrids, leading to increased yields by extending the crop season by 10-15 days. This is achieved by reducing the time spent on tillage operations. The implementation of zero-till maize in locations like coastal Andhra Pradesh, the Cauvery River delta (Tamil Nadu), and Karnataka following the rice harvest has shown to be a successful case of conservation agriculture (CA) in India. In these areas, farmers are able to achieve a yield of above 10 tonnes per hectare (Jat *et. al.*, 2009, Timsina *et. al.*, 2009). The implementation of Conservation Agriculture (CA) not only helps to save moisture and enhance soil health, but it has also demonstrated advantages in other cropping systems that involve the rotation of maize with wheat and mustard (Jat *et. al.*, 2011).

By using conservation agriculture methods like retaining residue and practicing crop rotations, soil erosion can be prevented and the amount of soil organic carbon can be increased. This ultimately leads to improved sustainability of maize ecosystems. Implementing site-specific nutrient management strategies alongside conservation agriculture can enhance both crop productivity and nutrients-use efficiency (Kumar et. al., 2015). Hence, the prospective implementation of Conservation Agriculture (CA) in conjunction with precision nutrient management during maize farming is anticipated to generate substantial advantages in India, augmenting both crop productivity and soil vitality.

RESULT AND DISCUSSIONS

Maize production in India has experienced significant growth, rising from 1.73 million metric tonnes in 1950-51 to 32 million metric tonnes in 2023. However, the current productivity level of 2.56 tonnes per hectare remains significantly lower than global averages, such as 10 tonnes per hectare in the United States. While the adoption of high-yielding hybrid seeds and modern farming techniques has contributed to productivity improvements, their implementation remains uneven across regions.

The demand for maize in India by 2030 is estimated at 50 million metric tonnes, driven by its diverse applications. Feed requirements, particularly in the poultry sector, are expected to account for 32 million metric tonnes. Industrial uses, including starch production and biofuels, will require an additional 25 million metric tonnes. Urban consumers' growing preference for specialty maize products like baby corn and popcorn will further increase demand, along with opportunities in export markets, especially in Southeast Asia and the Middle East.

Several challenges hinder the growth of maize production in India. Abiotic stresses, such as drought and heat stress, along with biotic stresses like pests and diseases, significantly impact yield stability. Inadequate post-harvest infrastructure exacerbates losses, particularly during storage and transportation. Furthermore, limited adoption of advanced production technologies, coupled with policy barriers such as delays in approving genetically modified maize, restricts potential advancements.

Ragni *et. al.*, 2024 observed that the plant extract also reduce the mycotoxin level in maize seeds & increase the seed viability for high yields to return the economical profit for grower in India.

Technological interventions offer promising solutions to address these challenges. The development of stress-resilient maize hybrids capable of withstanding climatic extremes is crucial. Conservation agriculture techniques, such as zero-tillage and crop residue retention, have shown potential in increasing yields while improving soil health. Additionally, advanced breeding methods, including the use of doubled haploidy and molecular markers, can accelerate the development of superior maize varieties. Policy reforms are equally essential for enhancing maize production. Expanding the adoption of single-cross hybrids, which offer significantly higher yields, and fostering public-private partnerships in seed production can bridge existing gaps. Establishing a supportive regulatory framework for genetically modified maize will provide farmers with additional tools to combat weeds and pests effectively. Improving post-harvest infrastructure, such as community-based dryers and metallic silos, can reduce losses and enhance grain quality. Finally, market stabilization measures and incentives for maize-based industries are necessary to ensure profitability for farmers and sustainability for the maize sector.

CONCLUSION

1. India's maize sector is at a critical juncture, with rising demand necessitating substantial improvements in productivity and sustainability. While the country has made commendable progress in maize production over the decades, significant gaps in yield and infrastructure remain.
2. Advanced breeding technologies, conservation agriculture, and stress-resilient hybrids hold the key to bridging the gaps.
3. Simultaneously, policy reforms focusing on hybrid seed adoption, genetically modified maize, and market support are vital.
4. Collaborative efforts between government agencies, private enterprises, and farmers can create an enabling environment for sustainable growth.

Overall, these challenges and opportunities, India can achieve its target of producing over 50 million metric tonnes of maize by 2030, contributing to food security, economic growth, and global agricultural competitiveness.

REFERENCES

1. Basappa, G., Deshmanya, J. B., & Patil, B. L. (2007). Post-harvest losses of maize crop in Karnataka: An economic analysis. Karnataka Journal of Agricultural Sciences, *20*, 69–71.
2. Bernstein, L., Bosch, P., Canziani, O., Chen, Z., Christ, R., Davidson, O., & Yohe, G. (2007). Climate change 2007: Synthesis report. Contribution of working groups I, II, and III to the Fourth Assessment Report of the Intergovernmental Panel on Climate Change. Intergovernmental Panel on Climate Change.
3. Blanco, M., & Gardner, C. (2014). Germplasm enhancement of maize: Strategies and synergy with maize curation. Illinois Corn Breeders School Proceedings, 1–2.

4. Cairns, J. E., Crossa, C., Zaidi, P. H., Grudloyma, P., Sanchez, C., Araus, J. L., Thaitad, S., Makumbi, D., Magorokosho, C., Bänziger, M., Menkir, A., Hearne, S., & Atlin, G. N. (2013). Identification of drought, heat, and combined drought and heat tolerance donors in maize (*Zea mays L.*). Crop Science, 53*(4)*, 1335–1340.

5. Castiglioni, P., Warner, D., Bensen, R. J., Anstrom, D. C., Harrison, J., Stoecker, M., Abad, M., Kumar, G., Salvador, S., D'Ordine, R., Navarro, S., Back, S., Fernandes, M., Targolli, J., Dasgupta, S., Bonin, C., Luethy, M. H., & Heard, J. E. (2008). Bacterial RNA chaperones confer abiotic stress tolerance in plants and improved grain yield in maize under water-limited conditions. Plant Physiology, 147*(2)*, 446–455.

6. Dhillon, B. S., Malhi, N. S., & Saxena, V. K. (1997). Development and improvement of heterotic populations in maize. CIMMYT Book of Abstracts: The Genetics and exploitation of heterosis in crops. An International Symposium, Mexico, DF, Mexico.

7. Diksha Sultania, G.Prasad and Lucky Rani (2022). Prevalence of aflatoxin in mize grains of Madhubani District-N-Bihar. Res. Jr. of *Agi. Sci.* 13 (4) : 1311–1313.

8. Eliasson, A. C. (2014). Starch in food: structure, function, and applications. CRC Press.

9. FAOSTAT. (2014). Production-Crops (2014) data. Food and Agriculture Organization of the United Nations.

10. Fischer, R. A., Byerlee, D., & Edmeades, G. O. (2014). Crop yields and global food security: Will yield increases continue to feed the world? ACIAR Monograph No. 158, Australian Centre for International Agricultural Research.

11. Jat, M. L., Dass, S., Sreelatha, D., Sai Kumar, R., Shekhar, J. C., & Chandana, P. (2009). Corn revolution in Andhra Pradesh: The role of single cross hybrids and zero-tillage technology. DMR Technical Bulletin 2009/5. Directorate of Maize Research.

12. Jat, S. L., Parihar, C. M., Singh, A. K., Jat, M. L., Jat, R. K., Singh, D. K., & Kumar, R. S. (2011). Conservation agriculture in maize production systems. DMR Technical Bulletin 2011/4. Directorate of Maize Research.

13. Kumar, R., Srinivas, K., & Sivaramane, N. (2013). Assessment of the maize situation, outlook, and investment opportunities in India. Country Report–Regional Assessment Asia (MAIZE-CRP), National Academy of Agricultural Research Management.

14. Kumar, V., Singh, A. K., Jat, S. L., Parihar, C. M., Pooniya, V., Singh, B., & Sharma, S. (2015). Precision nutrient and conservation agriculture practices for enhancing productivity, profitability, nutrient-use efficiencies, and soil

nutrient status of maize (*Zea mays L.*) hybrids. *Indian Journal of Agricultural Sciences*, 85, 926–930.

15. Kumari, Ragni, K. Salahuddin and G. Prasad (2024). Assessment of Andrographis Pariculata Plant extracts to reduce Mycotoxin Contamination in Maize Seeds, *KAVAKA 60*(3): 46–54.
16. Lucky Rani, D. Sultania, K Salahuddin, G. Prasad (2023) : Assessment of fungal diversity and impacts of aflotoxin B_1 on physiology of maize seeds (*Zea Mays L.*), J. of *Postharvest Technology*, 11*(3)* : 119–124.
17. NSSO. (2014). Employment and unemployment and household consumer expenditure (July 2011-June 2012), vol. *554* of the 68th round. National Sample Survey Office, Ministry of Statistics and Programme Implementation, Government of India.
18. Prasanna, B. M., Vivek, B. S., Sadananda, A. R., Jeffers, D., Zaidi, P. H., Boeber, C., Erenstein, O., Babu, R., Sudha, K. N., Gerard, B., Jat, M. L., Natalia, P., & Kevin, P. (2014). 12th Asian Maize Conference and Expert Consultation on Maize for Food, Feed, Nutrition, and Environmental Security, Bangkok, Thailand.
19. Prasanna, B. M. (2012). Diversity in global maize germplasm: characterization and utilization. *Journal of Biosciences*, 37(5), 843–855.
20. Prasanna, B. M., Araus, J. L., Crossa, J., Cairns, J. E., Palacios, N., Das, B., & Magorokosho, C. (2013). High-throughput and precision phenotyping for cereal breeding programs. In Cereal Genomics II (pp. 341-374). Springer.
21. Reddy, K. V. S., & Zehr, U. B. (2004). Novel strategies for overcoming pests and diseases in India. In new directions for a diverse planet. Proceedings of the 4th International Crop Science Congress, Brisbane, Australia, 26 Sep–1 Oct.
22. Timsina, J., Jat, M. L., & Majudar, K. (2009). Rice-maize systems of South Asia: Current status, future prospects, and research priorities for nutrient management. *Plant and Soil, 335*, 65–82.
23. Yadav, O. P., Hossain, F., Karjagi, C. G., Kumar, B., Zaidi, P. H., Jat, S. L., Chawla, J. S., Kaul, J., Hooda, K. S., Kumar, P., Yadava, P., & Dhillon, B. S. (2015). Genetic improvement of maize in India: retrospect and prospects. *Agricultural Research*, 4, 325-338.

24

Transformation of Agri-food Systems in Nepal

Shakeb Nabi

Country Director, Welt Hunger Hilfe, Nepal
Former Country Director, Bangladesh

Food security is one of the crucial issues which shape the well-being of a country. It is also one of the indicators which measures the well-being of the people globally. Food security in most cases is tantamount to national security, there is a strong correlation between the two. The world at this juncture is going through innumerable conflicts and humanitarian crises. It is argued that food security is one of the prime reasons for this conflict and it is vice versa as well, the prime reason for most of the conflicts is food insecurity and food crisis.

The world is going through a climate crisis with loss of biodiversity, frequent and unpredictable floods, river erosion, cyclones, droughts and GLOF at an increasing rate. There is already an established relationship between unsustainable food systems leading to climate change and vice-versa climate change leading to unsustainable food systems. Our current food system has failed us on achieving the triple bottle line People, Planet and Profit which is core to sustainable development. Our food systems have not been doing much good to people in providing healthy diets for all. Various studies have proven that the increase in non-communicable diseases like cancer, obesity, and respiratory problems is attributed to the quality of food we eat. The planet aspect of the triple bottom line, related to the environment, is having various negative impacts including the loss of bio-diversity, desertification, and deteriorating soil conditions. The last but not the least, profit is not benefitting the small and marginal farmers who are the major producers of food across the world. The situation are not different in Nepal as well.

Nepal is one of the countries where food security is a prominent concern. Only 48.2 per cent of households meet the minimum level of food security and 21 per cent of the population do not have enough food to eat, partly due to increasing food prices. Most regions in high hills and mountains in Provinces 6 and 7 experience high food insecurity (FAO 2022). The Nepal Annual Household Survey 2015/16 (CBS, 2016) reported that more than 80 per cent (by value) of food for household consumption was purchased in markets, with only 15 per cent from home production. One of the factors is the vulnerability of food systems to external shocks: violent conflicts, extreme weather events like drought or floods often exacerbated by climate change, market fluctuations and disruptions, and pandemics.

Declining youth interest in agriculture, migration from rural areas, changes in consumption patterns, the dependency on external food supplies even by rural producers, and a general shift to capital-intensive and globalized models of agriculture are challenges for the future of food production and rural livelihoods. One-third of the food produced is lost during production and transport or wasted by retailers and consumers. Key factors here include the lack of infrastructure facilities and poor storage and transportation facilities.

Getting obsessed with chemical fertilizers; the government of Nepal has allocated around 26% of the total agricultural budget for the subsidy of imported chemical fertilizers (Ministry of Finance, 2020). High dependence on neighbouring countries for inputs like chemical fertilizers and pesticides shows how fragile our food system is. If the supply chain is disrupted for any political reason or as for the recent COVID-19 lockdown, it causes massive food insecurity in the regions. So, it becomes apparent to make our agricultural system more resilient to external shocks

Food systems encompass the various activities, people, and places that relate to the production, processing, distribution, preparation, and consumption of food, distribution, preparation and consumption of food.

In the last few years, the concept of a Circular Economy has gained relevance on a global scale for its capacity to shift the current extractive linear economic model to a more sustainable circular model. It is argued that one of the methods that could solve the food security issues in high hills and mountains which are also food deficient is Circular Food Systems. Circular food systems are a set of practices and principles that reform food production, distribution and consumption. Governments, the private sector, and donors should promote locally and regionally anchored food systems more strongly that place a focus on the needs of both local small-scale farmers and poor consumers.

Some of the innovative examples from the work promoting circular food systems are the use of food and agriculture waste for producing bio-fertilizers, the use of native plants for bio-pesticides and recycling water from household chores for agriculture.

The circular economy comes at the rescue to address environmental issues simultaneously with economic development. By circular, an economy is envisaged as having no net effect on the environment; rather it restores any damage done in resource acquisition while ensuring little waste is generated throughout the process. A circular economy (CE) is known as a "closed-loop" economy, in which the industries aim for a regenerative system in which waste and input energy are minimized (Kyriakopoulos et al., 2019). A circular economy is based on the principles of designing out waste and pollution, keeping products and materials in use, and regenerating natural systems (Ellen MacArthur Foundation, 2017). So, the circular economy says that waste becomes a resource. Products are designed to be reused and recycled. Non-renewable natural resources are replaced by renewable. Energy is produced by renewable energy sources. People and industries are sharing items and services instead of ownership. The general objective of this study is to assess the concept of circular economy and its prospects in Nepalese agriculture. The specific objectives of the study are: Assessing the opportunities of circular economy in Nepalese agriculture. Linking circular agricultural approaches to global challenges.

Welthungerhilfe is an International NGO with the larger mandate of addressing issues around food and nutrition security with the larger mandate of a World Free from Hunger. The figure below demonstrates the Theory of Change applied by Welthungerhilfe globally to address hunger and poverty.

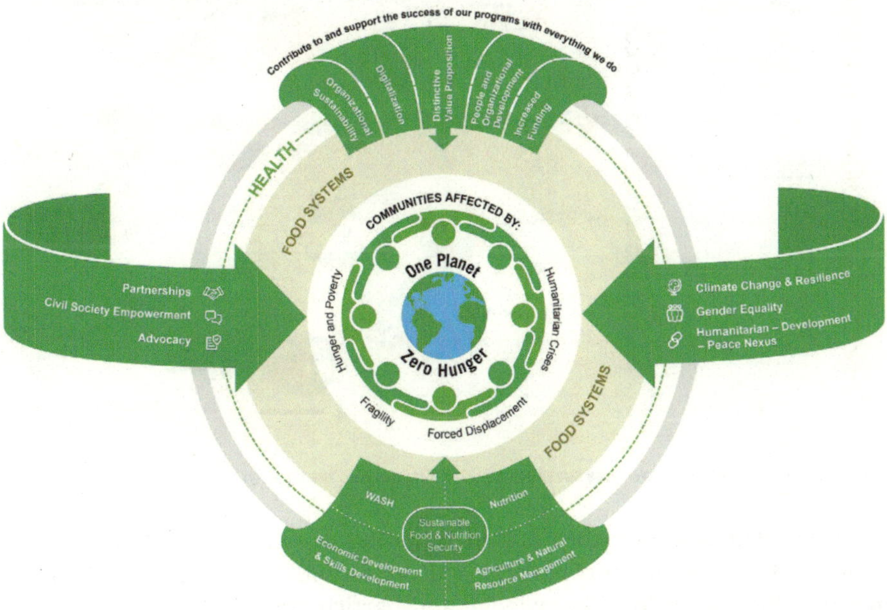

Food systems transformation through the agroecological approach is the key mandate of Welthungerhilfe. It has been addressing food insecurity and malnutrition in the food insecure pockets of Nepal mostly on the high hills and mountains which are hard to reach as well. WHH aims to achieve Zero Hunger through a systems transformation approach with four overarching principles as explained in the figure above. The first principle encompasses key intervention areas like Agriculture and Natural Resources Management, Economic Development & Skills Development, WASH and Nutrition directly addressing food and nutrition security at the community level. The second pillar is around partnerships with multiple stakeholders directly engaged in the food systems or has the potential to impact food systems. Advocacy with various stakeholders including the government and the private sectors are key ingredients of the strategy.

The third pillar of the WHH approach is more focused on addressing the impact of climate change on food security, ensuring that the transformed food system is equitable, and the fourth pillar is around the use of digital technology, institution development and ensuring sufficient funding for the same.

Welthungerhilfe approach to Food Systems Transformation is guided by the 13 principles of agro-ecology developed by HLPE-FSN(High Level Panel Experts on Food Security and Nutrition). HLPE-FSN is a United Nations body for assessing science related with world food security and nutrition. A brief graphic of the principle is depicted below.

THE FIVE LEVELS OF TRANSITION TOWARDS SUSTAINABLE FOOD SYSTEMS AND THE RELATED 13 PRINCIPLES OF AGROECOLOGY
SOURCE: GLIESSMAN (2007) AND HLPE (2019)

Agroecological approaches optimize the use of natural processes, limit the use of purchased inputs, promote closed cycles with very limited externalities and stress the use of local knowledge and participatory processes. It develops knowledge and practice through "learning by doing". It is one of the key approaches for promoting Sustainable, equitable, healthy and resilient food systems. It keeps the needs and aspirations of small and marginal actors in the food systems at its core and emphasizes on collective but differentiated responsibilities of various actors engaged in the food system.

ESG (environment, sustainability and governance) is the overall ambition associated with the transformation process. It is an action that uses nature to address environmental, social and economic challenges and helps in improving well-being and biodiversity.

Looking at the Nepal context where the incidence of food insecurity and malnutrition is higher in the high hills and mountains, WHH use the circular food systems approach focusing on the Locally Produced and Locally Consumed model. This model is also helping the revival of traditional food items, bringing diversity and nutritious food to the community food basket.

The circular food systems model, with the key overarching principle of People, Planet and Profit (Triple Bottomline) ensures that the transformative food systems are pro-people, pro-planet and ensure income to the small and marginalized groups associated with the food systems. It is in a way having a positive impact on Human Health, Economic Health and Environmental Health thus bringing an overall positive change.

Circular economy' in food and agriculture centers on the production of agricultural commodities using a minimal number of external inputs, closing nutrient loops and reducing negative discharges to the environment (in the form of wastes and emissions). This can be achieved through the (re)design of maintenance, repair, reuse, remanufacturing, refurbishing, and recycling (Bhattarai et al. 2020).

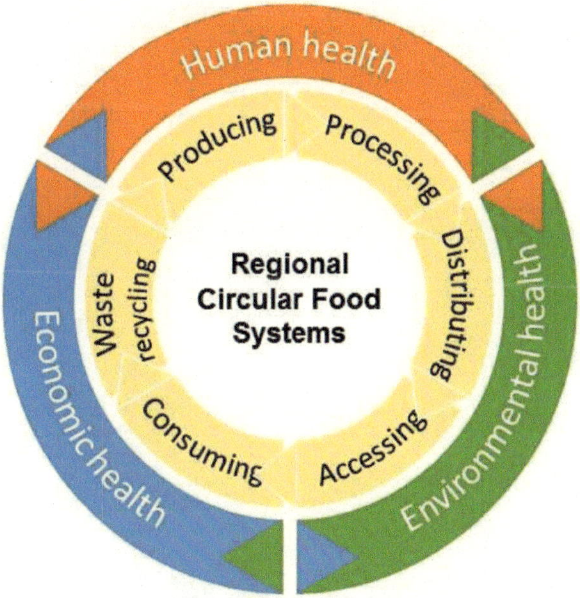

The four-food system is based on four key principles viz. extending the product life cycle, closing the loop (making it circular from linear), designing the efficiency and rethinking the system. It focuses on reducing, reusing, recycling and recovering, ensuring that there is no or optimum waste generated throughout the process. It also helps in climate mitigation by eliminating the use of chemicals in the food systems, reducing transportation, and food loss in transition. By focusing on the local food systems, WHH has also been helping the food system create more jobs, and save money, through the use of local resources (inputs).

It is argued that a circular food system could promote a resilient and sustainable food system, help generate productive employment, contain migration, revive the rural economy, and ensure food and nutrition security in hard-to-reach areas. Governments need to **strengthen the implementation of circular food economies** which are based on the continuous reuse of materials and products and the elimination of waste. Urban and peri-urban food systems need to be part of the overall food system transformations and innovations.